John Charles Snowball

The Cambridge Course of Elementary Natural Philosophy

Being the propositions in mechanics and hydrostatics in which those persons who are not candidates for honours are examined for the degree of B.A.

John Charles Snowball

The Cambridge Course of Elementary Natural Philosophy
Being the propositions in mechanics and hydrostatics in which those persons who are not candidates for honours are examined for the degree of B.A.

ISBN/EAN: 9783337192211

Printed in Europe, USA, Canada, Australia, Japan

Cover: Foto ©Thomas Meinert / pixelio.de

More available books at **www.hansebooks.com**

THE
CAMBRIDGE COURSE

OF

ELEMENTARY NATURAL PHILOSOPHY:

BEING THE PROPOSITIONS IN

MECHANICS AND HYDROSTATICS

IN WHICH THOSE PERSONS WHO ARE NOT CANDIDATES FOR HONOURS ARE
EXAMINED FOR THE DEGREE OF B.A.

ORIGINALLY COMPILED BY

J. C. SNOWBALL, M.A.

LATE FELLOW OF ST. JOHN'S COLLEGE, CAMBRIDGE.

FIFTH EDITION, REVISED AND ENLARGED,
AND ADAPTED FOR THE MIDDLE CLASS EXAMINATIONS.

BY

THOMAS LUND, B.D.

LATE FELLOW AND LECTURER OF ST. JOHN'S COLLEGE,
EDITOR OF WOOD'S ALGEBRA, &c.

Cambridge and London:
MACMILLAN AND CO.
1864.

ADVERTISEMENT.

The following work, originally compiled, at my suggestion, by the late Dr Snowball, has been unconditionally transferred to me by his Executors; and I have therefore not hesitated to re-write certain portions of it, and to make such alterations throughout as I conceived necessary. At the end of each chapter I have added a series of easy questions for the exercise of the Student. I have also carefully looked through *all* the papers of the University Examinations for the last ten years; and have printed a Collection of the best of the Problems, *with their Solutions*, as well as a large number of others *with Answers* only.

I am disposed to hope that the work, as it now appears, will be found suited to the wants, not only of University Students, but also of many others who require a short course of *Mechanics* and *Hydrostatics*, and especially of the Candidates at our Middle-Class Examinations.

T. L.

Morton Rectory, near Alfreton,
May 6, 1864.

CONTENTS.

MECHANICS.

Chapter I.

ARTS.
1—8. Definition of *Force, Weight, Quantity of Matter, Density, Measure* of Force.
10. *Pressure* another name for *Force* in Statics.
11. Definitions with respect to the action of Forces.
12. Forces properly represented by geometrical straight lines.
    Questions on Chap. I.

Chapter II.—*The Lever.*

17. Definitions of a Plane, a Solid, Parallel Planes, a Prism, and a Cylinder.
18. Definition of *Lever*.
19. Axioms.
20. Prop. I. A horizontal prism, or cylinder, of uniform density will produce the same effect by its weight as if it were collected at its middle point.
21. Prop. II. If two weights, acting perpendicularly on a straight *Lever* on opposite sides of the fulcrum, balance each other, they are inversely as their distances from the fulcrum; and the pressure on the fulcrum is equal to their sum.
22. Converse of Prop. II.
23. Prop. III. If two forces, acting perpendicularly on a straight *Lever* in opposite directions and on the same side of the fulcrum, balance each other, they are inversely as their distances from the fulcrum; and the pressure on the fulcrum is equal to the difference of the forces.

ARTS.
25. PROP. IV. To explain the different kinds of *Levers*.
26. PROP. V. If two forces, acting perpendicularly at the extremities of the arms of any *Lever*, balance each other, they are inversely as the arms.
27. PROP. VI. If two forces, acting at any angles on the arms of any *Lever*, balance each other, they are inversely as the perpendiculars drawn from the fulcrum to the directions in which the forces act.
28. Converse of Prop. VI.
29. PROP. VII. If two weights balance each other on a straight *Lever* when it is horizontal, they will balance each other in every position of the *Lever*.

Questions on Chap. II.

CHAPTER III.—*Composition and Resolution of Forces.*

30. Definition of Component and Resultant Forces.
31. PROP. VIII. If the adjacent sides of a parallelogram represent the component forces in direction and magnitude, the diagonal will represent the resultant force in direction and magnitude.
32. PROP. IX. If three forces, represented in magnitude and direction by the sides of a triangle, act on a point, they will keep it at rest.

Questions on Chap. III.

CHAPTER IV.—*Mechanical Powers.*

36. Definition of *Wheel-and-Axle*.
37. PROP. X. There is an equilibrium upon the *Wheel-and-Axle*, when the power is to the weight as the radius of the axle to the radius of the wheel.
40. Definition of *Pulley*.
41. PROP. XI. In the single moveable pulley, where the strings are parallel, there is an equilibrium when the power is to the weight as 1 to 2.

ARTS.
42. PROP. XII. In a system in which the same string passes round any number of pulleys and the parts of it between the pulleys are parallel, there is an equilibrium when Power (P) : Weight (W) :: 1 : the number of strings at the lower block.

43. PROP. XIII. In a system in which each pulley hangs by a separate string and the strings are parallel, there is an equilibrium when $P : W :: 1$: that power of 2 whose index is the number of moveable pulleys.

45. PROP. XIV. The Weight (W) being on an *Inclined Plane*, and the force (P) acting parallel to the plane, there is an equilibrium when $P : W ::$ the height of the plane : its length.

46. Definition of *Velocity*.

47. PROP. XV. Assuming that the arcs which subtend equal angles at the centres of two circles are as the radii of the circles, to shew that, if P and W balance each other on the *Wheel-and-Axle*, and the whole be put in motion, $P : W :: W$'s velocity : P's velocity.

48. PROP. XVI. To shew that if P and W balance each other on the machines described in Propositions XI, XII, XIII, and XIV, and the whole be put in motion, $P : W :: W$'s velocity in the direction of gravity : P's velocity.

Questions on Chap. IV.

CHAPTER V.—*The Centre of Gravity.*

49. Definition of *Centre of Gravity*.

50. PROP. XVII. If a body balance itself on a line in all positions, the centre of gravity is in that line.

51. PROP. XVIII. To find the centre of gravity of two heavy points; and to shew that the pressure at the centre of gravity is equal to the sum of the weights in all positions.

52. PROP. XIX. To find the centre of gravity of any number of heavy points; and to shew that the pressure at the centre of gravity is equal to the sum of the weights in all positions.

54. PROP. XX. To find the centre of gravity of a straight line.

55. PROP. XXI. To find the centre of gravity of a triangle.

ARTS.

56. PROP. XXII. When a body is placed on a horizontal plane, it will stand or fall, according as the vertical line, drawn from its centre of gravity, falls within or without its base.

57. PROP. XXIII. When a body is suspended from a point, it will rest with its centre of gravity in the vertical line passing through the point of suspension.

Questions on Chap. V.

HYDROSTATICS.

CHAPTER I.

59. Definitions of Fluid; of elastic and non-elastic Fluids.

CHAPTER II.—*Pressure of non-elastic Fluids.*

60. PROP. I. Fluids press equally in all directions.

61. PROP. II. The pressure upon any particle of a fluid of uniform density is proportional to its depth below the surface of the fluid.

62. PROP. III. The surface of every fluid at rest is horizontal.

63. PROP. IV. If a vessel, the bottom of which is horizontal and the sides vertical, be filled with fluid, the pressure upon the bottom will be equal to the weight of the fluid.

64. The pressure of a fluid on any horizontal plane placed in it, is equal to the weight of a column of the fluid whose base is the area of the plane, and whose height is the depth of the plane below the horizontal surface of the fluid.

66. PROP. V. To explain the *Hydrostatic Paradox*.

67. PROP. VI. If a body floats on a fluid, it displaces as much of the fluid as is equal in weight to the weight of the body; and it presses downwards, and is pressed upwards, with a force equal to the weight of the fluid displaced.

Questions on Chaps. I. and II.

Chapter III.—*Specific Gravities.*

ARTS.
69. Definition of Specific Gravity.
70. Prop. VII. If M be the magnitude of a body, S its specific gravity, and W its weight, $W = MS$.
71, 72. To find the relation which exists between the weights, magnitudes, and specific gravities, of two substances, and of a compound formed of them.
73. Prop. VIII. When a body of uniform density floats on a fluid, the part immersed : the whole body :: the specific gravity of the body : the specific gravity of the fluid.
74. Prop. IX. When a body is immersed in a fluid, the weight lost : whole weight of the body :: the specific gravity of the fluid : the specific gravity of the body.
76. Prop. X. To describe the *Hydrostatic Balance;* and to shew how to find the specific gravity of a body by means of it, 1st, when its specific gravity is greater than that of the fluid in which it is weighed; 2ndly, when it is less.
77. Prop. XI. To describe the common *Hydrometer;* and to shew how to compare the specific gravities of two fluids by means of it.

Questions on Chap. III.

Chapter IV.—*Elastic Fluids.*

79. Prop. XII. Air has weight.
81. Prop. XIII. The elastic force of air at a given temperature varies as the density.
82. Prop. XIV. The elastic force of air is increased by an increase of temperature.
84. Prop. XV. To describe the construction of the *Common Air-Pump*, and its operation.
85. Prop. XVI. To describe the construction of the *Condenser*, and its operation.

CONTENTS.

ARTS.
86. PROP. XVII. To explain the construction of the *Common Barometer;* and to shew that the mercury is sustained in it by the pressure of the air on the surface of the mercury in the basin.

87. PROP. XVIII. The pressure of the atmosphere is accurately measured by the weight of the column of mercury in the *Barometer*.

92. PROP. XIX. To describe the construction of the *Common Pump*, and its operation.

93. PROP. XX. To describe the construction of the *Forcing-Pump*, and its operation.

94. PROP. XXI. To explain the action of the *Siphon*.

96. PROP. XXII. To shew how to graduate a common *Thermometer*.

97. PROP. XXIII. Having given the number of degrees on *Fahrenheit's* thermometer, to find the corresponding number on the *Centigrade* thermometer.

Questions on Chap. IV.

EXAMPLES AND PROBLEMS WITH SOLUTIONS.
EXAMPLES AND PROBLEMS WITH ANSWERS.
UNIVERSITY EXAMINATION PAPERS.

MECHANICS.

[*The explanatory matter, printed in small type, forms no actual part of the* UNIVERSITY COURSE; *but is illustrative of the particular Definition, or Proposition, which it immediately follows; and will be found useful for answering the Questions, and solving the Problems, which are usually given in the University Examinations.*]

CHAPTER I.

DEFINITIONS; EXPLANATION OF STATICAL FORCES; THE MANNER IN WHICH THEY ARE MEASURED, AND REPRESENTED.

1. MECHANICS is the science, which treats of the causes that prevent, or produce, *motion* in bodies, or that *tend* to prevent, or produce, *motion*.

It is divided into two parts. The one, which investigates the conditions fulfilled when a body is in a state of *rest*, is called STATICS. The other, which treats of the causes and the effects of *motion*, is called DYNAMICS.

Thus it is the province of STATICS to shew how the roof of a building is supported by the beams and the walls. If the roof gave way, and fell, it would belong to DYNAMICS to account for the circumstances attending the fall—to explain why the *motion* took place in one direction rather than in another—to determine the time elapsed in falling—and the swiftness of the motion at any instant.

The part of MECHANICS treated on in the following pages is STATICS.

L. C. C.   1

2. *Definition of* Force.

Whatever the cause be which produces, or prevents, motion, or which tends to produce, or to prevent, motion, in a body, it is called a Force.

If a heavy body, as a stone, be laid on the open hand, experience shews that, to prevent the stone from falling, the hand must make some effort. Again, to set a ball rolling along the ground requires some exertion. The effort, or exertion, is called in either case a force; and although the effect produced be not great enough to prevent entirely the fall of the stone, or to communicate motion to the ball, yet it is still called a force.

From the definition of Statics given in Art. 1, it will readily be understood, that in that branch of Mechanics the conditions are investigated which are fulfilled by those *Forces* only, which keep a body *at rest*.

3. *Definition of* Weight.

All bodies, if left to themselves, fall, or tend to fall, towards the earth's centre, through a power, which resides in the earth, of constantly drawing all substances towards it, called the force of gravity. Consequently, if any body be reduced to a state of *rest*, it exerts a certain pressure downwards upon that which sustains it. And the *precise amount* of this pressure for any particular body is called the Weight of that body.

4. The weights of different bodies may be *compared* thus:—Let two bodies be successively attached in the same manner to a spring, so that they may act upon it by their *weights* in the same way. If they produce the same effects, (by bending the spring to the same extent,) the *weights* of the bodies are equal. Any other body, which produces the same effect on the spring by its *weight* as both the former bodies when applied together do by their *weights*, has its weight *double* of that of either of them. And by means of such a contrivance as this spring, bodies might be shewn to be three, four, or any number of, times the *weight* of a given body.

5. The weight of any body is *measured* thus:—The *weight* of a certain bulk of some particular substance is first fixed upon as a *standard*. Thus the weight of a piece of *lead* of a certain size being called a *pound*, any other body, which by the force of gravity only, produces the same effect as four, or six, or ten, such pieces of

lead, will be four, or six, or ten, *pounds* in *weight*, as the case may be.

6. *Definition of* QUANTITY OF MATTER.

The substance, material, or stuff, of which any body is made, is called MATTER. And since all bodies have *Weight*, the property of having *Weight* is to be considered as necessarily belonging to *Matter*. Hence in the same ratio, or degree, that one body has more *weight* than another, it is concluded, that it contains more *matter;* that is, the *Quantity of Matter in a body is proportional to its Weight.*

Thus, if a body A weigh *one* pound, and another body B weigh *three* pounds, the *quantity of matter* contained in A is *said* to be to the quantity contained in B as 1 to 3; or B is said to contain three times as much *matter* as A does.

7. The exact *quantity of matter* contained in any body may be *measured* by comparing its *weight* with the *weight* of some particular body, which has been fixed upon for a *standard*. Thus, if a *cubic inch* of *water* be previously taken as the body by which to measure the *quantities of matter* contained in all other bodies, and the *quantity of matter* in this cubic inch of water be called 1, then the *quantity of matter* in any other body would properly be said to be 5, if the *weight* of that body were *five times* as great as the *weight* of the cubic inch of water.

8. *Definition of* DENSITY.

The *Density* of a substance, or body, is the degree of closeness, with which the matter composing it is, as it were, packed; which closeness is *measured*, or *compared* in manner following:—

Let *equal* bulks of two different substances be taken, suppose Water and Lead. Then, if the bulk of *water* which is taken weigh *one* pound, it will be found, that the piece of lead of equal size with it will weigh $11\frac{4}{10}$ pounds. There is evidently, therefore, $11\frac{4}{10}$ times as much heavy matter in a piece of lead as there is in an *equal bulk* of water; and this fact is expressed, or described, by saying, that "The DENSITY of Lead is to the DENSITY of Water, as $11\frac{4}{10}$ is to 1"; or by saying, "The DENSITY of Lead is $11\frac{4}{10}$ times that of Water."

If the *Density* of water be *called* 1, that is, if *water* be taken as a standard, to *measure Density*, then the *Density* of lead will be properly called $11\frac{4}{10}$, or 11·4.

In the same manner as it has been explained how the *Density* of *lead* is estimated with respect to the *Density* of *water*, the *Densities* of any other substances, whether solid or fluid, may be determined with respect to that of *water*.

9. *Definition of* "Measure of Force".

In *Statics* a Force is *measured* by the *weight* which it would support. In other words, the amount of a Statical *Force* is expressed by stating the *number of pounds* it would support, if the *Force* were made to act directly opposite to the *Force of Gravity*.

Thus, if the *weight* of a body were P pounds, and it were prevented from moving towards the earth's surface by a hand placed beneath it, the resistance offered by the hand to the communication of motion (that is, the *force* exerted by the hand), would be P pounds; and if this same pressure were produced by the hand *in any other direction*, it would be described in the same manner, by saying that it was "*equal to P pounds*", or that it *was* "*P pounds*". If, therefore, a *Force* be represented by P, it is meant that P is the *number of pounds* which the *Force* would support, on the supposition that the *Force* is made to act directly opposite to the *Force of Gravity*. In other words, P is the *number of pounds*, which the *Force* is just able to lift.

Forces, *in* Statics, *also called* Pressures.

10. In whatever *direction* a *Force* tends to produce motion, its magnitude, as has already been stated, is *measured* by the *weight* of the body which would exert the same effect to produce motion *downwards*, as the *Force* under consideration exerts in the line in which *it* endeavours to produce motion. And that such a method of measuring *Forces* is allowable appears from *this* consideration, viz., that the effect produced by the *weight* of a heavy body* may be made to take place in *any* direction whatever; horizontally, as in the case of a string being attached to an object lying on a table and kept stretched by a heavy body (W), which hangs over the edge o

* By '*a heavy body*', in Mechanics, is simply meant a body acted on by the *Force of Gravity*.

the table, as in fig. (1); or vertically upwards, by passing the string over a peg A, and attaching the end to a ring B, so that BA may be vertical, as in fig. (2); or *in any other direction*, as in fig. (3), by making the heavy body pull the string in the

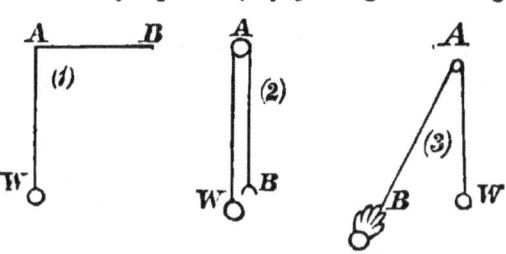

line BA, which is inclined at *any* angle to that (AW) in which it acts itself.

11. *Definitions with respect to the* action *of* FORCES.

(1) The *point* at which a *Force* acts upon a body is called the "*point of application*" of the *Force*.

(2) The *line* in which a *Force*, acting alone, produces, or tends to produce, motion, is called "*the line of the Force's action*"; and *any* line which is parallel to the line of a *Force's* action is said to be "*in the direction of the Force's action*", or "*in the direction of the Force*".

When the *direction of a Force's action*, (or, as it is generally called, "*the direction of the Force*",) is indicated by a *line*, either the very line is given in which the *Force* acts, or some line which is parallel to it. "*The line of a Force's action*", and "*the direction of the Force*", must by no means be confounded together. If the former be known, the latter is necessarily known also; but if only the latter be given, the precise line in which the *Force* produces, or tends to produce, motion, is uncertain; and all that can be said respecting it is, that the line of action of the *Force* is either that given line, or some other line which is *parallel* to it.

(3) If two or more *Forces* be applied to a body, or at some point, and no motion is produced, they are said to "*counteract*", or to "*balance*", one another, or to be "*in equilibrium*".

12. *Forces properly represented by geometrical straight lines.*

Since *lines* may be drawn of any length, and in any direction, from a point, the *lines* in which *Forces* act, and the ratios which the *Forces* bear to one another, may be represented by drawing *lines*, which coincide with the *lines* in which the *Forces* act, and whose lengths bear to one another the same ratios that the *Forces* themselves bear to one another.

Among other advantages which attend this method of expressing the magnitudes and directions of *Forces*, the *addition* and *subtraction* of such *Forces* as act at a point *in the same straight line* are easily effected. Thus, if a certain *Force* act at A in the line AH, and AB be taken to represent it, and another *Force*, half as great as the former, act at A in the same direction, and also tend to move the body from A towards H, then, by taking BC equal to the half of AB, the line AC will represent the whole pressure at A, both with respect to the *magnitude* of that pressure, and to *the line in which it acts*.

And, in like manner, if a *Force* equal to *half* the original *Force* AB act at A in the line AH, but tends to move the body at A from A towards K, half the pressure of the former *Force* will be counteracted by this new *Force*. Cutting off from the line AB, therefore, a part BD equal to the half of AB, the effective pressure still remaining will be properly represented by AD, with respect both to its *magnitude*, and to its *line of action*.

13. N.B. It will be gathered from the above, that a *Force AB* applied to A has not the same effect as a force BA applied at that point; for a *Force AB* would tend to move a body at A in the line KH towards H, but a *Force BA* would tend to move a body at A in the line KH from A towards K. It is not, therefore, indifferent whether the words "a *Force AB*", or "a *Force BA*", be used; since, though the two *Forces* represented by AB, and BA, are the same in *magnitude*, and also act in the same straight line, yet they tend to produce motions directly *opposite* to one another, the *Force AB* tending to move the body at A towards H, and the force BA tending to move the body at A towards K.

14. *The effect produced at a point by any Force is the same at whatever point in its line of action the Force is applied, provided the latter point be supposed rigidly connected with the former.*

Thus, if a body P be suspended by a string CP, the *Force* necessary to prevent P falling to the earth is found to be the same whether that *Force* be applied at A, or B, or C;—the weight of the *string* being either neglected, or the weight of that portion of it which is supported along with the heavy body P, being counterbalanced. And although, in this case, the points A, B, C, are not, in fact, *rigidly* connected with one another, and with P, the result is the same as if they were, but in certain other cases the *rigidity* of the system is necessary.

15. DEF. If a string, fastened at one end, be pulled by a *Force* applied at the other end, the resistance to motion made by the string at any point in it is called the TENSION of the string at that point.

If the string be supposed to be without weight, it will follow, from Art. 14, that the *tension* at *every* point of it is the same; namely, the *Force* by which the string is pulled.

16. *To recapitulate the substance of this Chapter.*

(1) The precise amount of the *Force*, or pressure, with which any particular body endeavours to move towards the earth, is called the WEIGHT of that body. Art. 3.

This WEIGHT of a body is *measured* by comparing the tendency of the body to move towards the earth with that of some other given body (of a certain size and formed of a certain material) which is taken for a standard, and to which the name of a *grain*, an *ounce*, a *pound*, or a *ton* is given, as the case may be. Art. 4.

(2) MATTER is the substance of which all bodies are composed. It is found, by universal experience, to have a tendency to move towards the earth. Art. 6.

(3) THE QUANTITIES OF MATTER contained in different bodies (whether the bodies be great or small, rare like gas, or dense like lead) are *proportional*, (not *equal*,) to the weights of the bodies. Art. 6.

(4) The DENSITIES of different substances are *proportional*, (not *equal*,) to the weights of *equal bulks* of the substances. Art. 8.

(5) FORCE. I. Whatever be the cause which moves, or tends to move, matter, existing under any form whatsoever, it is called a FORCE.

FORCES which prevent motion taking place,—that is, STATICAL FORCES,—are measured by the number of *pounds* they would support if they acted vertically upwards. Arts. 2, 9.

II. THE LINE OF A FORCE'S ACTION is the actual line in which the force tends to produce, or to prevent, motion. Art. 11. And if a *Force* be said to be P, it is meant that it is equal to P *pounds*.

III. THE DIRECTION OF A FORCE is indicated either by the line of its action, or by any line which is *parallel* to the line of its action. Art. 11.

IV. The *magnitudes* of *Forces*, and the lines (or the *directions*) in which they act, may be represented by means of straight lines. Art. 12.

V. If a straight line AB represent a *Force* acting on a material point placed at A, and BA represent another *Force* acting on the same point, the two *Forces* AB and BA are equal, and tend to move the point in the same straight line, but in opposite ways from A. Art. 13.

VI. A Statical *Force* produces the same effect at whatever point in its line of action it is applied. Art. 14.

VII. To investigate the effects produced by a *Force*, there must be given:—

1st. The *magnitude* of the *Force;*—which is known, if the number of pounds be known which the *Force* would support.

2nd. The *point* at which the *Force* is *applied.*

3rd. The *line* in which the *Force* acts,—or the *direction;* for knowing the *direction* of the *Force*, and the point it is applied at, the *line* in which the *Force* acts may be determined by drawing a line through the given point parallel to the given direction.

Questions on Chap. I.

(1) What are the two great divisions of *Mechanics* called, and how are they separately distinguished?

(2) What is *Force?* Give an example.

(3) Define *Weight,* and explain how it is reduced to *numbers,* so as to become the subject of calculation.

(4) Is the *Weight* of the same body at the same place invariable? Is the *Weight* of a body changed by changing its *figure?*

(5) How do you define *Matter?* and *Quantity of Matter?*

(6) How is the *Quantity of Matter* in a body ascertained and *measured?*

(7) Which has the greater *Quantity of Matter*, a feather-bed weighing 50 lbs., or a child of the same weight?

(8) What is *Density?* How is it reduced to *numbers,* and measured?

(9) If the *weights* of equal *bulks* of two substances are in the ratio 3 : 1, what is the ratio of their *Densities?*

(10) Is the *Density* of the same body at the same place invariable, like *Weight?* Give an illustration.

(11) How is FORCE *measured* in Statics? Does your *measure* of *Force* apply to the case of a *Force* acting *vertically upwards?* also to a *Force* acting in *any* direction whatever?

(12) Before we can estimate the effect of a *Force* upon a body, what three things are required to be known?

(13) Can a *Force* be represented by a straight line in *magnitude*, as well as *direction?*

(14) If it be stated that two forces of 5 lbs., and 10 lbs., act upon a body, what more is wanting to enable us to determine the result?

(15) Is the "*direction* of the *Force*", and the "*line of action* of the *Force*", the same thing?

(16) If a *Force* be applied to a body by means of a *string,* the weight of which is inconsiderable, is the result affected by the *length* of the *string?*

CHAPTER II.

THE LEVER.

17. DEFS. (1) A PLANE, OR A PLANE SUPERFICIES, is that superficies, or surface, in which, if *any two* points be taken, the straight line joining them lies wholly within the superficies. EUCLID, I. Def. 7.

(2) A SOLID is that which hath length, breadth, and thickness. EUCLID, XI. Def. 1.

(3) PARALLEL PLANES are such as do not meet one another, however far they may be produced. EUCLID, XI. Def. 8.

(4) A PRISM is a solid bounded by plane rectilineal figures, of which two that are opposite are equal, similar, and parallel to one another; and the others all parallelograms. EUCLID, XI. Def. 13. It is further necessary, that the parallel figures have their equal angles opposite *each to each.*

Thus, let *ABCD*, and *abcd*, be two equal and similar quadrilateral figures, placed with their planes parallel, and with their similar angles opposite, each to each; and let all the figures, such as *ABba*, which are formed by joining the equal angles of *ABCD* and *abcd*, be parallelograms; then the solid included by these parallelograms, and by the *ends*, or *bases*, *ABCD* and *abcd*, is called a *prism*.

The *length* of the prism is any one of the edges *Aa*, *Bb*, &c.; which lines, being the sides of adjacent parallelograms, are all equal to one another.

If the parallelograms be *rectangles*, the solid is a *rectangular prism*.

(5) A CYLINDER is a solid described by a rectangle, *ABCD*, revolving round one of its sides, *AB*, which remains fixed. EUCLID, XI. Def. 21.

The side *AB*, which remains fixed, is the *length* of the *cylinder*, and is called its *axis*.

The surfaces described by the two sides, *AD*, and *BC*, which are adjacent to the fixed side of the rectangle, are circles, and are parallel to each other; they are called the *ends*, or *bases*, of the *cylinder*.

18. *Definition of* LEVER.

A LEVER is a rigid rod, moveable in one plane round a fixed point in it called the FULCRUM.

The two parts into which the rod is divided by the *fulcrum* are called the *Arms* of the *Lever*.

In the following *Propositions* the thickness of the rod is neglected, and the *Lever* is considered to be a *geometrical line, without weight*, but still rigid and inflexible. The *Weights*, or *Forces*, acting on the *Lever*, are supposed to act *in the same plane*.

19. The properties of the *Lever* are sometimes deduced from the following principles, the truth of which will be readily admitted, and which are therefore called *Axioms*:—

AXIOM I. Equal forces, acting perpendicularly at the extremities of equal arms of a *Lever*, exert the same effort to turn the *Lever* round.

THE LEVER.

AXIOM II. If two weights balance each other upon a straight *Lever*, the pressure upon the fulcrum is equal to the sum of the weights, whatever be the length of the *Lever*.

AXIOM III. If a weight be supported upon a *Lever*, which rests on two fulcrums, the pressures on the fulcrums are together equal to the whole weight.

Axiom I. is quite self-evident, since the forces are equal, and act upon the *Lever* in a manner perfectly similar.

In Axiom II., the weights act in the same, i.e. in a vertical, direction, and therefore cannot *in any degree* counteract each other. Hence it is obvious, that the effective force on the *Lever* must be the *sum* of the weights, and this is supported by the fulcrum, since there is equilibrium. Therefore the pressure on the fulcrum is equal to the sum of the weights.

In Axiom III., there is nothing to support the weight but the fulcrums; therefore the pressure on the fulcrums must be *together* equal to the whole weight.

20. PROP. I. *A horizontal prism, or cylinder, of uniform density, will produce the same effect by its weight, as if it were collected at its middle point.*

Let AB be the *axis* of the given prism, or cylinder, supposed to be held at rest in a horizontal position; C its middle point; DCE the vertical line through C meeting the outer surface of the prism, or cylinder, in D and E.

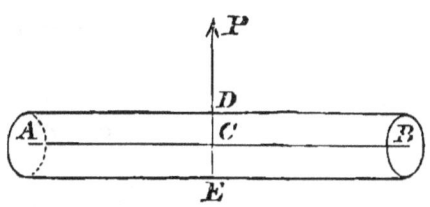

Affix a string to the prism, or cylinder, at D; and apply a force P vertically upwards equal to W, the weight of the prism, or cylinder. Then, if the prism, or cylinder, be left free to move, by withdrawing the forces which first held it in its horizontal position, it is evident, that no motion will ensue. For, since P is *exactly equal* to W, by supposition, both which forces act upon the prism, or cylinder, in exactly opposite directions, there can be no motion *upwards* or *downwards*. Nor can there be any *angular* motion of the prism, or cylinder. round D, because the parts AD, BD, being exactly similar and equal, no reason can be assigned for an *angular* motion in one direction, which would not be equally valid for an *angular* motion in the opposite direction, and therefore there will be no *angular* motion at all.

Since, then, the *whole* effect of the prism, or cylinder, to produce motion by its *weight* is exactly counteracted by a single force P, equal to W, acting in the line ECD, that effect is exactly equivalent to a single force W acting at *any point* in the same line, and therefore at C the middle point (Art. 14).

Cor. 1. Hence it follows, that any uniform *rod, in a horizontal position*, produces the same effect by its weight to turn the rod round any fulcrum in it, as if its whole weight were concentrated in its middle point, the rigidity of the rod being supposed to be still maintained.

Cor. 2. Hence also it follows, that any horizontal prism, or cylinder, of uniform density, will *balance* on its middle point; and the pressure on a fulcrum placed there will be the weight of the prism, or cylinder. And conversely, if such a prism, or cylinder, balance on a point in itself, that point is either its *middle* point, or in the vertical line which passes through its *middle* point.

21. Prop. II. *If two weights, acting perpendicularly on a straight Lever, on opposite sides of the fulcrum, balance each other, they are inversely as their distances from the fulcrum; and the pressure on the fulcrum is equal to their sum.*

Let the two weights P, Q, acting perpendicularly at M and N on the straight *Lever* MCN, whose fulcrum is C, balance each other, the *Lever* being in a horizontal position. It is required to shew, that $P : Q :: CN : CM$.

In MN take a point E such, that
$$ME : MN :: P : P+Q, \text{ Euclid, vi. 12,}$$
then $ME : MN-ME :: P : P+Q-P$,
or $ME : NE :: P : Q$.

Produce MN both ways to A and B, making MA equal to ME, and NB equal to NE. Then, since
$$ME : NE :: P : Q,$$
$$2ME : 2NE :: P : Q,$$
or $\quad AE : BE :: P : Q.$

Now suppose AB to be a uniform rod, whose weight is $P+Q$. Then the weights of AE, and BE, will be P and Q, respectively. And since P and Q, acting at the middle points of the *lines* AE and BE, balance on C, therefore the portions of the rod, AE and BE, which may be substituted for P and Q by Prop. I., will also balance round C, that is, the whole rod AB balances itself on C, and therefore C is its *middle point*.

But $P : Q :: AE : BE,$
$$:: AB - BE : AB - AE,$$
$$:: 2BC - 2BN : 2AC - 2AM,$$
$$:: BC - BN : AC - AM,$$
$$:: CN : CM.$$

Also the Pressure on the fulcrum is not altered by substituting the rod for the weights, and therefore

$\qquad =$ whole weight of the rod,
$\qquad = P + Q$, the sum of the weights.

22. Conversely, *if two weights, or forces, acting perpendicularly on a straight Lever, on opposite sides of the fulcrum, are inversely as their distances from the fulcrum, they will balance each other.*

To prove this, construct as in the foregoing Proposition, and assume that $P : Q :: CN : CM$; then from this it will be required to shew, that C is *the middle point* of the rod AB, and therefore that the rod will *balance* on C, by Cor. 2, Prop. I.

The proof will stand thus,
$$ME : NE :: P : Q,$$
$$:: CN : CM,$$
$$\therefore ME + NE : ME :: CN + CM : CN,$$
or $MN : ME :: MN : CN, \therefore ME = CN.$
Also $MN : NE :: MN : CM, \therefore NE = CM.$

But, by construction, $MA = ME$, and $NB = NE$;

$$\therefore AC = MA + CM = ME + NE = MN,$$
$$\text{and } BC = NB + CN = NE + ME = MN;$$
$$\therefore AC = BC, \text{ and } \therefore \text{ the rod } AB \text{ will } balance \text{ on } C.$$

Then, since the weight of $AE = P$, and weight of $BE = Q$, by Prop. I., P and Q will also balance on the *Lever MCN*.

23. PROP. III. *If two forces, acting perpendicularly on a straight Lever in opposite directions and on the same side of the fulcrum, balance each other, they are inversely as their distances from the fulcrum; and the pressure on the fulcrum is equal to the difference of the forces.*

Let the two *Forces* P and Q, acting perpendicularly at M and N, on the straight *Lever MC*, in opposite directions, and on the same side of the fulcrum C, balance each other. Then it is to be shewn, that $P : Q :: CN : CM$; and (Q being the *Force* which is the nearer to the fulcrum) that the pressure on the fulcrum $= Q - P$.

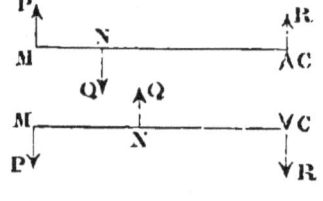

Suppose the fulcrum at C removed, and let its resistance (R) be supplied by a *Force* equal to R, and acting perpendicularly to the *Lever* in the same direction as P. The equilibrium will not be disturbed.

Then since P and R are exactly counterbalanced by Q, they must produce a pressure at N equal and opposite to Q. Let Q be removed, and its place supplied by a *fulcrum* on the contrary side of the *Lever* to that on which Q acted, sustaining the pressure (namely Q) produced by P and R. The equilibrium is still maintained; and the case is now that of two *Forces* acting perpendicularly on *opposite* sides of the fulcrum, and balancing each other; and therefore (by Prop. II.),

$$P : R :: CN : NM;$$
$$\therefore P : P + R :: CN : CN + NM :: CN : CM.$$

But Q, the pressure on the fulcrum which has been supposed to be placed at N, is equal to $P + R$, by Prop. II.;

$$\therefore P : Q :: CN : CM.$$

Also since $Q = P + R$, $\therefore R = Q - P$,
that is, the pressure on the fulcrum = the difference of the forces.

24. From the last two Propositions it appears, that if a straight *Lever*, which is acted on *perpendicularly* by two weights, or other *Forces*, P and Q, respectively applied at the distances CM and CN from the *fulcrum* C, be at rest, then, whether P and Q act on the same side, or on different sides, of the *fulcrum*, the proportion $P : Q :: CN : CM$ is always true.

Hence also $P \times CM = Q \times CN$ (*Wood's Algebra*, Art. 237) is an *equation* which expresses the conditions of equilibrium in *all* such cases.

There is no impropriety in multiplying a *Force* by a *line*, because both are expressed in *numbers*, when they become subjects of calculation. Thus a *force* of 3 lbs. acting perpendicularly on a straight lever at a distance of 4 feet from the fulcrum will balance another *force* of 6 lbs. acting at a distance of 2 feet on the opposite side of the fulcrum and in the same direction, because $3 \times 4 = 12 = 6 \times 2$.

The product $P \times CM$ is sometimes called *the moment of P about C*; and, similarly, $Q \times CN$ is *the moment of Q about C*. Hence, in the last two Propositions, the *moments of P and Q are equal*.

Also, since if $P : Q :: CN : CM$, it is proved that P and Q will balance on C, therefore, *conversely*, if *the moments* of P and Q with respect to C are equal, they will balance each other.

When the *Lever* is used to balance a given *Force*, Q, by the application of another *Force*, P, Q is usually called "*the Weight*", and P "*the Power*".

If CM, the perpendicular distance from the fulcrum at which the *Power* acts, be greater than CN, the distance at which the *Weight* acts, the *Power* required to balance the *Weight* is *less than* the *Weight;* in this case "*force*" is said to be "*gained*" by the application of the *Lever*. But if CM be less than CN, the *Power* required to balance the *Weight* is greater than the *Weight*, and "*force*" is then said to be "*lost*".

25. PROP. IV. *To explain the different kinds of* LEVERS.

LEVERS are divided into three classes, according to the relative position of the points where the *Power* and the *Weight* are applied with respect to the *Fulcrum*.

(1) Where the *Power* (*P*) and the *Weight* (*Q*) act on *opposite* sides of the *Fulcrum* (*C*), as thus

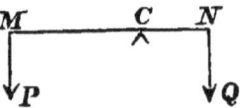

(2) Where the *Power* and the *Weight* act on the *same* side of the *Fulcrum*, but the perpendicular distance from the *Fulcrum* at which the *Power* acts is *greater* than that at which the *Weight* acts, as thus

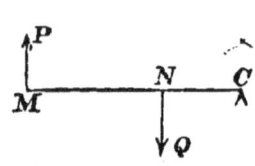

(3) Where the *Power* and the *Weight* act on the *same* side of the *Fulcrum*, but the perpendicular distance from the *Fulcrum* at which the *Power* acts is *less* than that at which the *Weight* acts, as thus

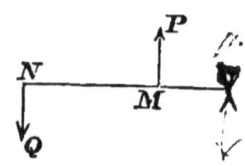

Of the FIRST class the poker, when used to raise the coals, is an instance; the bar of the grate on which the poker rests being the *Fulcrum*, the force exerted by the hand the *Power*, and the resistance of the coals the *Weight*. In the common *Balance*, the *Power* and the *Weight* are equal *Forces* perpendicularly applied at the ends of *equal arms*. In the *Steelyard*, the *Power* and the *Weight* are perpendicularly applied at the ends of *unequal arms*. Pincers, scissors, and snuffers, are double *Levers* of this kind, the rivet being the *Fulcrum*.

Since *CM* may be either greater or less than *CN*, the *Power* in *Levers* of this class may be either less, or greater, than the *Weight*, and consequently "*Force*" may be either "*gained*", or "*lost*", by using them.

Of the SECOND class, a *cutting blade*, such as is used by coopers, moveable round one end, which is fastened by a staple to a block, and worked by means of a handle fixed at the other end, is an example. An *oar* is also such a *Lever;* the *Fulcrum* being the extremity of the blade (which remains fixed, or nearly so, during the stroke), the muscular strength and weight of the rower being the *Power*, and the *Weight* being the resistance of the water to the motion of the boat, which is counteracted and overcome at the rowlock. A pair of *nutcrackers* also is a double *Lever* of the second class.

Here, since *CM* is greater than *CN*, the *Power* is always less than the *Weight*, or *Force is "gained"* by using *Levers* of the *second* class.

An example of the THIRD class is the board which the turner (or knifegrinder) presses with his foot to put the wheel of his lathe in motion; the *Fulcrum* being the end of the board which rests on the ground, the *Power* being the pressure of the foot, and the *Weight* being the pressure produced at the crank put on the axletree of the wheel. *Fire-tongs* and *sugar-tongs* are double *Levers* of this kind; the *Weight* in either case being the resistance of the substance grasped. The limbs of animals are also such *Levers :* thus, if a weight be held in the hand and the arm be raised round the elbow as a *Fulcrum*, the *Weight* is supported by muscles fastened at one extremity to the upper arm, and again attached to the fore-arm, after passing through a kind of loop at the elbow.

Here, since *CM* is less than *CN*, the *Power* is greater than the *Weight*, or *Force is "lost"* by making use of *Levers* of the *third* class.

26. PROP. V. *If two forces, acting perpendicularly at the extremities of the arms of any Lever, balance each other, they are inversely as the arms.*

In this Prop. the *arms* of the Lever are supposed *straight,* but joined together *at the fulcrum* so as not to be in *the same* straight line.

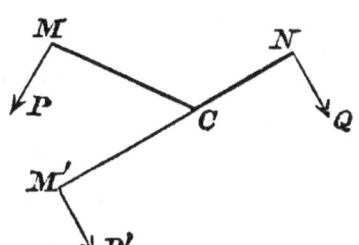

Let the two forces P and Q, acting perpendicularly at the extremities of the straight arms, CM and CN, of any *Lever* whose fulcrum is C, balance each other; then

$P : Q :: CN : CM.$

For, suppose the arm NC produced to M', so that $CM' = CM$; and suppose a force P', equal to P, to act perpendicularly at M' on the

L. C. C. 2

Lever M'CN; then, since $P = P'$, and $CM' = CM$, by Axiom I., Art. 19, the effort of P to turn the *Lever MCN* round C is equal to that of P' on the *Lever M'CN*. But, by the supposition, P balances Q on the *Lever MCN*; therefore also P' balances Q on the *straight Lever M'CN*.

Hence, as before proved in Prop. II,
$P' : Q :: CN : CM'$; but $P = P'$, and $CM = CM'$,
$\therefore P : Q :: CN : CM$.

Cor. Here again, as in the two preceding Propositions,
$$P \times CM = Q \times CN.$$

27. Prop. VI. *If two forces, acting at any angles on the arms of any Lever, balance each other, they are inversely as the perpendiculars drawn from the fulcrum to the directions*[*] *in which the forces act.*

Let P and Q be two forces, which, acting at any angles on the arms CA and CB of any Lever ACB, balance each other about the fulcrum C; and let the perpendiculars CM and CN be drawn from the fulcrum C to the lines in which the forces act; then

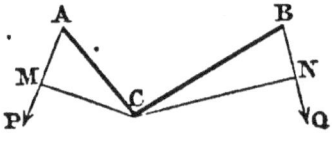

$P : Q :: CN : CM$.

For, since a force produces the same effect at whatever point in its line of action it is applied (Art. 14), the force P may be supposed to be applied at M; and in order that it *may* be so applied, let a rod, CMA, supposed without weight, be fastened to CA. In like manner, Q may be supposed to be applied at N perpendicularly to the part CN of the rod CNB which is added to CB.

And, since P acting at M *perpendicularly* to CM balances Q acting at N *perpendicularly* to CN, $\therefore$, by Prop. V,
$P : Q :: CN : CM$;

and therefore also, when P and Q, acting at A and B in the lines AP, BQ, balance, $P : Q :: CN : CM$.

---

[*] More correctly, "to the *lines of action* of the forces". See Art. 11.

28. COR. Conversely, if P and Q be inversely as the perpendiculars from the fulcrum upon their lines of action, they will *balance* each other.

For suppose Q' to be the force which applied at B in BN balances P at A; then, by what has been proved,
$$P : Q' :: CN : CM.$$
But, by supposition, $P : Q :: CN : CM$;
$$\therefore P : Q' :: P : Q, \text{ or } Q' = Q;$$
but Q' balances P, $\therefore Q$ also balances P.

In this Proposition the *arms* of the Lever may be either straight or crooked, since nothing in the proof is made to depend upon the particular *form* of CA, or CB. The *rigidity*, however, of the *arms*, whatever be their form, is a *necessary* condition.

It may also be noticed here, that every possible case of two forces balancing on a *Lever* has now been discussed. In Prop. II. the *Lever* is *straight*, and the forces act *perpendicularly* to the arms *on opposite sides* of the fulcrum. Prop. III. is the same as Prop. II., except that the forces act *on the same side* of the fulcrum. In Prop. V. the forces still act *perpendicularly* to the arms, but the *Lever* is *bent*. In Prop. VI. the forces act *at any angles* to the arms, and the *Lever* is either *straight* or *bent*. But *in every case*
$$P \times CM = Q \times CN;$$
from which *equation*, any three of the quantities being given, the fourth may be found.

29. PROP. VII. *If two weights balance each other on a straight Lever when it is horizontal, they will balance each other in every position of the Lever.*

Let P and Q be two weights, which balance each other round the fulcrum C on the straight Lever ACB, when it is *horizontal*. They will balance each other on the *Lever*, when it is made to take *any* other position, as $A'CB'$.

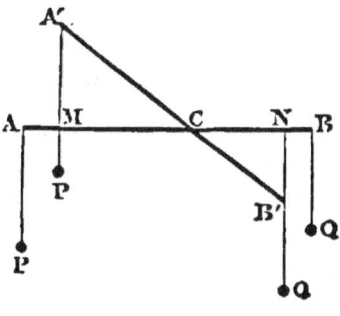

Produce QB' to cut AB in N, and $A'P$, if necessary, to cut AC in M.

Since weights act perpendicularly to the horizon, $A'P$ and $QB'N$ are both perpendicular to the horizontal line ACB;

$\therefore$ angle CMA' = right angle = angle CNB',

and angle $A'CM$ = opposite angle $B'CN$, Euc. I. 15.

$\therefore$ also angle $CA'M$ = angle $CB'N$, and the triangles $CA'M$, $CB'N$, are equiangular, and therefore *similar*. Hence, by Euc. VI. 4,

$$CN : CB' :: CM : CA';$$

alternando, $CN : CM :: CB' : CA',$

$$:: CB : CA.$$

But, since P and Q balance on ACB, $CB : CA :: P : Q$;

$$\therefore CN : CM :: P : Q.$$

But CM and CN are the perpendiculars from C on the lines in which P and Q act, when they are hung at A' and B'; therefore, by Prop. VI., Cor., P and Q will *balance* on $A'CB'$; and since $A'CB'$ is the *Lever* in *any* position, the above proof applies to *every* position of the *Lever*.

Cor. 1. Hence, if two weights do *not* balance each other on a straight *Lever*, when the *Lever* is horizontal, they cannot balance each other in an inclined position of the *Lever*.

For if they *did* balance in an inclined position, it would follow from Prop. VI., that

$$P : Q :: CN : CM,$$

$\therefore P : Q :: CB' : CA'$, by what has been proved,

$$:: CB : CA;$$

and $\therefore P$ and Q balance in the *horizontal* position of the lever, which is contrary to the supposition.

Cor. 2. Hence, also, if two *weights*, acting freely, balance each other on a straight *Lever* in *any one* position of the *Lever*, except the vertical, they will balance in *every other* position of the *Lever*.

For the *ratio* $CN : CM$ is independent of the angle at which the *Lever* is inclined; therefore if it *once* satisfies the conditions of equilibrium, it will do so *always*.

Questions on Chap. II.

> (1) What is a *Lever?* Is there any such *Lever* in practice as that which is assumed in this chapter?

> (2) What is the *fulcrum*, and what are the *arms*, of a *Lever?* Must the *arms* necessarily lie on opposite sides of the *fulcrum?*

(3) In *Axiom* II., if the *Lever* itself be supposed to have *weight*, how will the result be affected?

(4) In *Axiom* III., if the weight be placed exactly *half-way* between the fulcrums, what is the pressure on *each?*

(5) In *Prop.* I., if the prism, or cylinder, were not *horizontal*, how would the proof be affected?

(6) In Prop. I., if the prism, or cylinder, were *not* of *uniform density*, how would the proof be affected?

(7) What is the meaning of "*effect*" in the enunciation of *Prop.* I.

(8) In *Prop.* II., would P and Q balance, if they were to exchange places? Are there any *other* points between M and N at which they would balance?

(9) In *Prop.* II., if Q were doubled, where must P act to maintain the equilibrium?

> (10) If 2 cwt., acting at a distance from the fulcrum of 1 foot, is balanced on a horizontal straight *Lever* by a power of 28 lbs. acting perpendicularly, what is the length of *arm* at which the *power* acts?

(11) In *Prop.* III., where P and Q balance each other, acting on the same side of the fulcrum, would the equilibrium be disturbed, if P were doubled, and CM halved? Also would the pressure on the fulcrum remain the same?

> (12) There are *three* classes of *Levers;* what is it which distinguishes one class from another?

(13) Is power "*lost*" or "*gained*" in the use of *fire-tongs?*

(14) Where would you place the nut in a pair of nut-crackers to produce the greatest effect; and why?

(15) How does the contrivance of placing the row-locks *outside* the boat affect the efforts of the rower?

(16) In Prop. VI. is it necessary that the angles at which the forces act should be *equal* to one another?

If the forces *once* balanced acting at equal angles, would the same forces balance on the same Lever acting at *any other* equal angles?

CHAPTER III.

COMPOSITION AND RESOLUTION OF FORCES.

> 30. *Definition of* COMPONENT *and* RESULTANT FORCES.

It is found, by experiment, that a body which is acted on by two forces applied, at the same instant and *in different lines*, to the same point of it, instead of moving, or having a tendency to move, in either of the lines in which the forces act, moves, or has a tendency to move, in a line lying between them. Whence it appears, that the two original forces by their combined action produce the effect of a single third force, which third force is called, from the circumstance of its *resulting* from the actions of the original forces, their "*Resultant*" with respect to them; while they are called, with respect to it, its "*Components*".

The *Resultant* (R), which produces the same effect as the compound action of the original forces P and Q applied at the same point at the same instant, is said to be "*compounded*" of P and Q. This *Resultant* (R) also, if conceived to be the sole original force, may be supposed to be "*resolved*" into the two forces P and Q; since those two forces, acting in the manner described (namely, at the same point, and at the same instant), produce exactly the same effect on the body as the single force R does.

Similarly, if there be *more than two* forces, acting at the same point, and at the same instant, the resulting action is found to be such as can be produced by a certain single force, which latter force is therefore called the *Resultant* of all the other forces, whilst those other forces are called the *Components* of such *Resultant*.

> 31. PROP. VIII. *If the adjacent sides of a parallelogram represent the* component *forces in direction and magnitude, the diagonal will represent the* resultant *force in direction and magnitude.*

COMPOSITION AND RESOLUTION OF FORCES. 23

Let AB and AC represent, in direction and magnitude, the two *component* forces which act at A. Complete the parallelogram $ACDB$, and draw the diagonal AD. Then AD will represent the *Resultant* of AB and AC, (1) in direction, and (2) in magnitude.

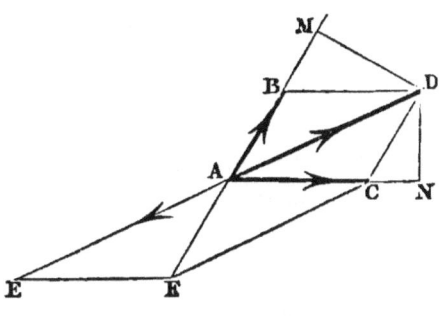

From D draw DM and DN perpendiculars to AB and AC, produced if necessary.

(1) Then in the triangles DBM, DCN,

$$\angle DMB = \text{a right angle} = \angle DNC,$$

and $\angle DBM = \angle BAC$, (since BD, AC are parallel, and MBA cuts them), $= \angle DCN$;

∴ the third angle, BDM, of the one triangle = the third angle, CDN, of the other; and the triangles are equiangular and similar;

hence, $CD : DN :: BD : DM$,

and *alternately*, $CD : BD :: DN : DM$.

Now, if there be a *Lever* AD whose fulcrum is D, which is acted on by the forces AB, AC, applied at A, since

Force in the line AM : force in the line AN

:: $AB : AC$,

:: $CD : BD$,

:: $DN : DM$,

the two forces acting on the *Lever* AD are inversely as the perpendiculars from the fulcrum on their lines of action, and therefore the *Lever* will be kept at rest about D by them (Art. 28). Wherefore the *Lever* will also be kept at rest by the *Resultant* of those forces; because that single force produces the same effect as they do, when they act at the same point and at the same instant.

This *Resultant* therefore must act *in the line AD*, for it keeps the *Lever* at rest, which it could not do, were it to act at A and make any angle with the *Lever AD* on either side of it.

Hence the *diagonal* of the parallelogram represents the *Resultant* in *direction*.

(2) Again: Having shewn, that the *Resultant* of AB and AC acts in the line of the *diagonal*, next to prove that the *diagonal* represents it in *magnitude* as well as in *direction*.

Produce DA, and suppose a force AE to be taken in it equal and opposite to the *Resultant* of AB and AC.

The joint effect of AB and AC will now be counteracted by AE; and the point A, which is acted on by the three forces AB, AC, and AE, will remain at rest, so that any one of them may be considered as the *Resultant* of the other two.

Whatever, therefore, be the effect produced by the joint action of AE and AC, it is counteracted by AB; that is, AB must be equal and opposite to the *Resultant* of AE and AC.

Complete the parallelogram $AEFC$, and draw the diagonal AF. By the first part of the Proposition, AF is the line in which the *Resultant* of AE and AC acts; and since the force AB is equal and opposite to that resultant, AF must be in the same straight line with AB, and, therefore, it is parallel to CD.

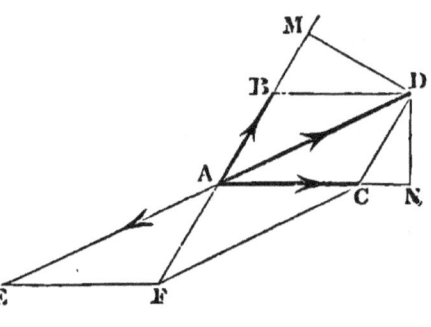

Hence $ADCF$ is a parallelogram; and therefore

$$AE = FC = AD.$$

The *Resultant*, therefore, of AB and AC (which is equal and opposite to AE) will be properly represented in

COMPOSITION AND RESOLUTION OF FORCES.   25

magnitude by AD, the diagonal of the parallelogram of which AB and AC are the sides.

This Theorem is commonly called "*The Parallelogram of Forces*".

> 32. PROP. IX. *If three forces, represented in magnitude and direction by the three sides of a triangle, act on a point, they will keep it at rest.*

Let the sides AB, BC, CA, *taken in order**, of the triangle ABC, represent in *magnitude* and *direction* (see Art. 11) three forces which act on the point A; they will keep A at rest.

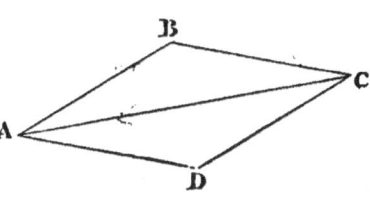

Complete the parallelogram $ABCD$.

Then AD is parallel and equal to BC; and it will, therefore, represent in *magnitude* and in *line of action* (Art. 11) the force, which acts at the point A in the *direction* BC and is represented in *magnitude* by BC.

Now the forces AB and AD acting at A will produce a *Resultant* AC, by Prop. VIII.

If, therefore, a force CA act at A, the force AC will be counteracted, and the point A will remain at rest. Wherefore, if three forces, represented in *magnitude* and *direction* by AB, AD, CA,—(or, which is the same thing, if they be represented by the three sides, AB, BC, CA, taken in order, of the triangle ABC)—act on the point A, they will keep it at rest.

33. COR. 1. It appears from this proof, that if two sides of a triangle ABC, as AB and BC, *taken in order*, represent in *magnitude* and *direction* two forces which act at the same instant on the point A, the third side of the triangle, AC, not taken in the same order as AB and BC, represents their *Resultant* in magnitude and direction.

* By the expression "*taken in order*" it is meant, that, if ABC be the triangle, and AB be one of the forces, BC (and not CB) is the next, and CA (not AC) is the third; so that the forces are described *in the same direction* round the triangle, proceeding from A to B, from B to C, and from C to A again.

34. COR. 2. Since the forces, which keep the point at rest, are represented by the sides of a *triangle*, it follows that the sum of any two of them must necessarily be greater than the third. EUCL. I. XX.

QUESTIONS ON CHAP. III.

(1) HAVE two or more *forces*, which act *in the same straight line*, a '*Resultant*'? If so, how is it determined?

(2) Can the *Resultant* of two forces in any case exceed the *sum* of the forces? Under what circumstances is it *least*? Can it ever be *nil*?

(3) If two forces, acting on a point at the same instant, are given both in magnitude and direction, their *Resultant* is readily found by Prop. VIII. Conversely, having given the *Resultant*, can you find the *Components*?

(4) The *Resultant* of *two* forces is generally determined by the construction of a parallelogram; how would you determine the *Resultant* of *three* or *more* forces?

(5) Can the *Resultant* in any case be *equal* to *one* of the *Components*? If so, what are the conditions?

(6) Two forces, represented by 3 and 4, act on a point in directions *at right angles* to each other, what is the numerical measure of the *Resultant*?

(7) In Prop. IX. how can three forces, represented in magnitude and direction by the three sides of a triangle, *act on a point*? Does *direction* always mean the *line of action*?

(8) Is it possible for three forces, represented by the numbers 3, 4, 7, acting on a point, to keep it at rest? If not, why not?

(9) At what angle must two *equal* forces act, that their *Resultant* may be equal to *each* of them?

(10) If p and q represent two forces, acting on a point in directions *at right angles* to each other, what is the algebraical expression for the *Resultant*?

(11) The *Resultant* of two forces, which act on a point at right angles to each other, is given both in magnitude and direction, and the *direction* of one of the forces is also known. From these data can you determine geometrically both *Components*?

(12) If *four* forces, represented by the sides of a square, *taken in order*, act on a point, trace out the effect, and say what it will be.

CHAPTER IV.

MECHANICAL POWERS.

> 35. The 'Mechanical Powers' are certain simple machines by means of which *power* is, for the most part, gained in the application of *force* either to support weights, or to give motion to bodies.

They are *six* in number, the *Lever*, the *Wheel and Axle*, the *Pulley*, the *Inclined Plane*, the *Wedge*, and the *Screw*.

The *Lever* has been already treated of in Chap. II. The *Wedge*, and the *Screw*, are not included in this Course. We proceed with the

WHEEL AND AXLE.

> 36. *Definition of* Wheel *and* Axle.

The Wheel and Axle consists of a cylinder AB, called the *Axle*, rigidly attached to another cylinder CD of greater diameter, called the *Wheel*. The two cylinders have a *common* axis, EF, about which the machine can turn. The *Power* (P) acts by means of a rope, or chain, coiled round the *Wheel*, and fastened to it at one end; and the *Weight* (W) acts in a similar manner by means of a rope, or chain, coiled about the *Axle*, but tending to turn the machine round in the *opposite* direction.

If the axis EF be supported in an *horizontal* position, then P and W may *both* be represented by *weights* hanging vertically, as in the annexed diagram.

37. Prop. X. *There is an equilibrium upon the Wheel and Axle, when the Power is to the Weight as the radius of the Axle to the radius of the Wheel.*

The efforts of P and W to turn the machine round its axis will be the same in whatever plane they act perpendicular to the axis.

Suppose, then, the lines in which the Power and the Weight act to be in *the* plane which coincides with the junction of the *Axle* with the *Wheel*. Since the *Wheel* and *Axle* have a common axis, this plane, or section, will present the appearance of two concentric circles, as AB, and CD, with a common centre O. Join O with M and N, the points in which the strings leave (and therefore touch) the circumferences of the circles.

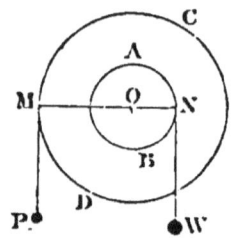

OM and ON are therefore perpendicular to MP and NW, the lines in which the *Power* and the *Weight* act, and the whole machine may be considered as a simple *Lever*, MON, with fulcrum O, and the two forces, P and W, acting perpendicularly at M and N.

Now there will be equilibrium, by Art. 28, in this *Lever*, and therefore in the case of the *Wheel* and *Axle*, when

$$P : W :: ON : OM,$$

:: radius of *Axle* : radius of *Wheel*.

38. COR. 1. When P and W are both weights, since OM and ON are both perpendicular to the vertical lines in which P and W act, and also pass through the same point O, MON is a *horizontal straight line*.

39. COR. 2. If the thickness of the ropes, by which P and W act, be taken into account, and each be represented by $2t$, then, since the action of the power and weight must be supposed to be transmitted along the *axes* of the ropes, there will be equilibrium, when

$$P : W :: ON+t : OM+t.$$

Hence it appears, that the ratio of the *Power* to the *Weight* is greater as the thickness of the ropes is increased; for, if any quantity be added to the terms of a ratio of less inequality, that ratio is increased. (Wood's *Algebra*, Art. 227.)

Wherefore, also, it is shewn, that the mechanical advantage in using the *Wheel* and *Axle* is *diminished* by increasing the thickness of the ropes or chains. For the ratio of P to W is increased, which means that a greater *power* is required to balance a given *weight*.

THE PULLEY.

> **40.** *Definition of* PULLEY.

A PULLEY is a small wheel moveable about an axis through its centre, and having a groove of uniform depth along its outer edge to admit a rope or flexible chain. The ends of the axis are fixed in a frame called the *Block*.

The *Pulley* is said to be *fixed*, or *moveable*, according as the *Block* is *fixed*, or *moveable*. The *Power* acts by the rope, or chain, or string, which works in the groove; and the *Weight* is fixed to the *Block*.

In the following Propositions the groove of the *Pulley* is supposed perfectly *smooth;* and the *weight* of the machine is not taken into account.

> **41.** PROP. XI. *In the single moveable pulley, where the strings are parallel, there is an equilibrium when the Power is to the Weight as* 1 *to* 2*.

The annexed diagram represents the case here supposed.

A power, P, acts by means of a string $PADBR$, which passes under the moveable pulley and is made fast at R. A weight, W, is fixed to the *block* by a string, and the line of its action is CW, C being the centre of the section of the pulley made by a plane passing through PA and RB. By the supposition, PA, RB, and CW, are *parallel;* and they are in one plane. Also PA, RB, are *tangents* to the circle at A and B; therefore, if AC, BC, be joined, $\angle PAC = $ a right angle $= \angle RBC$, and $\therefore$ ACB *is a straight line*, and CW is at right angles to AB.

Now, the whole machine being at rest under the action of these forces, we may consider it as a straight *lever* AB kept at rest round B, as a fulcrum, by the power P acting perpendicularly upwards at A, and the weight W acting

---

* More correctly thus :—"*when* there is equilibrium, the Power is to the Weight &c."

perpendicularly downwards at C; but in this case, and therefore in the case first supposed, the forces are inversely as their distances from the fulcrum, that is,

$$P : W :: BC : AB :: 1 : 2.$$

42. PROP. XII. *In a system in which the same string passes round any number of pulleys, and the parts of it between the pulleys are parallel, there is an equilibrium when Power* (P) : *Weight* (W) :: 1 : *the number of strings at the lower block**.

The annexed diagram represents such a case as is here supposed. The system consists of two blocks, having each *two* pulleys, the upper block being *fixed*, and the lower one *moveable*. The string, by which P acts passes round each of the pulleys, as shewn in the figure—the several portions of it are *parallel* to each other, and to the line AW in which the Weight (W) acts.

Since the same string continuously passes round all the pulleys, its tension must be everywhere the same, otherwise motion will ensue, which is contrary to the supposition. But for the *outer* portion of the string, to which the *power* is immediately applied, this tension is P; therefore it is P throughout; that is, *each* string at the lower block exerts a force P in opposition to W. And the forces are all *parallel*, therefore, since they balance, W is equal to the *sum* of those which act in the opposite direction, that is, is equal to P multiplied by the number of strings at the lower block, or $P : W :: 1 :$ N°. of strings at the lower block, whatever that number may be.

43. PROP. XIII. *In a system in which each pulley hangs by a separate string, and the strings are parallel,*

---

* More correctly thus :—"*when* there is equilibrium, the Power is to the Weight &c."

there is an equilibrium when $P : W :: 1 :$ *that power of 2 whose index is the number of moveable pulleys**.

In this system the strings are *fixed* at $F, G, H,$ &c. and pass round the moveable pulleys $A, B, C,$ &c. respectively, as in the figure, to the last of which the power P is applied. The strings are all parallel, and in the direction in which the weight W acts.

Now, C being a single moveable pulley with parallel strings, when there is equilibrium, by Prop. XI., the pressure downwards at $C = 2P$; therefore the Tension of the string from C round the pulley $B = 2P$.

Hence, the weight supported at the moveable pulley $B = 2 \times (2P) = 2^2 \times P =$ tension of string which passes round the pulley A.

So, weight supported at moveable pulley A
$= 2 \times (2^2 \times P) = 2^3 \times P$.

Therefore, when there is equilibrium on the system of *three* moveable pulleys, as here represented,

$$W = 2^3 \times P.$$

And so on, by the same mode of reasoning, if n be *any* number of moveable pulleys, it will appear, that, when there is equilibrium, $W = 2^n \times P$;

$$\therefore \frac{P}{W} = \frac{1}{2^n}; \quad \text{or, } P : W :: 1 : 2^n.$$

44. *Definition of* INCLINED PLANE.

An *Inclined Plane*, in Mechanics, means a plane *inclined* to the horizon, that is, neither horizontal, nor vertical.

The plane is here supposed to be perfectly smooth, and rigid, and capable of counteracting, and entirely destroying, the effect of any force which acts upon it *in a direction perpendicular to its surface.*

When, therefore, a body is sustained on an inclined plane, by a *Power* directly applied to it, the case is that of a body kept at rest

---

* More correctly thus :—"*when* there is equilibrium, $P : W$ &c."

by *three* forces, its own *Weight* acting vertically, the power applied, and the resistance of the plane in a direction at right angles to its surface. The *Power* must also obviously act in the *same plane* as the other two forces, otherwise motion would ensue.

It is evident, that the *Inclined Plane* will bear, or take off, a *portion* of the *Weight*, how much is the question; that is, we are required to find the ratio of P to W, when there is equilibrium.

45. Prop. XIV. *The weight (W) being on an Inclined Plane, and the force (P) acting parallel to the plane, there is an equilibrium when $P : W ::$ the height of the plane : its length*.*

[The '*length*' of the inclined plane is the *section* of it made by the plane in which the forces act—the '*height*' of the plane is the perpendicular let fall from the highest point in it to meet the horizontal plane through the lowest point—and the '*base*' is the distance from the lowest point to this perpendicular. Thus, in the annexed diagram, AB is the '*length*' of the plane, BC its '*height*', and AC its '*base*'.]

Let AB be the *length* of the *Inclined Plane*; AC its horizontal *base*; and BC, perpendicular to AC, its *height*. Let the Weight (W) be supported on the plane at D by the Power (P) acting in the direction DB parallel to the plane; then, when there is equilibrium,

$$P : W :: BC : AB.$$

From D draw DE at right angles to AB, meeting the base in E; and from E draw EF vertical, or at right angles to AE, meeting AB in F.

Then in the triangles EFD and ABC, ∵ FE, BC are parallel, angle $DFE =$ angle ABC;

and angle $FDE =$ a right angle $=$ angle BCA,

∴ angle $DEF =$ angle CAB, and ∴ the triangles are equiangular and similar. Hence $DF : FE :: BC : AB$.

* More correctly thus:—"*when there is equilibrium, $P : W$ &c.*"

Now the body at D is kept at rest by three forces, the weight W acting vertically, the reaction of the plane acting at right angles, and the power P acting parallel to the plane. And these forces are respectively parallel to the sides of the triangle DEF, therefore those sides will represent them in magnitude as well as in direction (by the converse of Prop. IX.). Hence

$$P : W :: DF : FE,$$
$$:: BC : AB, \text{ (by what has been proved.)}$$

46. *Definition of* VELOCITY.

By the VELOCITY of a body in motion is meant the Degree of Swiftness, or Speed, with which the body is moving. And this "degree of swiftness" is described, or *measured*, by stating how long the line is that the body moves through, with uniform swiftness, in some *given* portion of time.

Thus a clear notion would be conveyed of the *Velocity* of a coach, if it were said to be *nine miles* in *an hour;* the *space* moved over by the coach being *nine miles*, and *an hour* being the *portion of time* during which the motion took place. If only the *space* that is moved through were mentioned, and nothing were said about the *time* of describing that space,—or if the *time* only were given,—it is evident that no idea could be formed of *the degree of swiftness*, that is, of the *Velocity*, of the coach.

The *Velocity* of a body is *measured*, for the most part, in mathematical investigations, by the number of *feet* passed over by the body, moving uniformly, in a *second* of time.

COR. Since the quicker a body moves the more space it will pass over in a *given* time, it will follow from the observations just made, that *The Velocities of two bodies which move during any* (THE SAME) *time, are in the ratio of the spaces which the bodies respectively describe—each with uniform swiftness—in that time.*

47. PROP. XV. *Assuming that the arcs which subtend equal angles at the centres of two circles are as the radii of the circles, to shew that, if P and W balance each other on the Wheel and Axle, and the whole be put in motion,* $P : W :: W\text{'s velocity} : P\text{'s velocity}.$

L. C. C.

Suppose P and W to act (as in Prop. x.) at the circumferences of the Wheel and the Axle, in the same plane, perpendicularly to the horizontal radii OM, ON; and let the whole be put in motion round the axis*, so that mOn may become the horizontal diameter, instead of MON. P will now act at m at right angles to Om, and W will act at n at right angles to On; and the velocities of P and W will be as Mm to Nn, since Mm, being the length of string unwrapped from the wheel, is the space through which P will have descended in the same time that W has ascended through Nn.

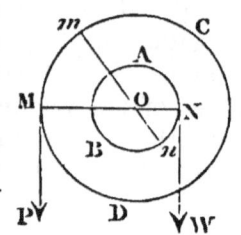

Now, since P and W balance each other, by the supposition,

$P : W :: ON : OM$, (by Prop. x,)

$:: Nn : Mm$, by the assumption,

$\therefore P : W :: W$'s velocity $: P$'s velocity.

Cor. $P \times P$'s velocity $= W \times W$'s velocity.

48. Prop. XVI. *To shew that if P and W balance each other on the Machines described in Propositions XI., XII., XIII., and XIV., and the whole be put in motion, $P : W :: W$'s velocity in the direction of gravity $: P$'s velocity.*

1st. In the case of the single moveable pulley (Prop. XI.), let C, the centre of the pulley (see Fig.) be raised through any height, as an inch; W will thereby be also raised through an inch, and each of the strings RB, AP, will have been shortened an inch; so that, if P continue to keep the string tight, it will have moved through two inches in the time that W has been raised one inch.

---

* This motion is supposed to be given by the application of some extraneous force, which is removed as soon as the displacement of the machine is effected; and then the system is *in equilibrium* in its new position.

But $P : W :: 1 : 2$, by Prop. XI., since they balance each other,

$$:: 1 \text{ inch} : 2 \text{ inches},$$

$$:: \text{space described by } W : \text{space described in the same time by } P;$$

$\therefore P : W :: W\text{'s velocity} : P\text{'s velocity}.$

2nd. In the system where the same string passes round all the pulleys, and the parts of it between the pulleys are parallel, as in Prop. XII. (see Fig.), if the lower block be raised through any height, as an inch, *each* of the strings between the upper and lower blocks will be *shortened* an inch, and therefore in the time that W moved through one inch, in order to have kept the string tight, P will have moved through as many inches as there are parallel strings at the lower block.

But $P : W :: 1 :$ No. of strings at lower block, since they balance each other,

$$:: \text{space described by } W : \text{space described in same time by } P,$$

$\therefore P : W :: W\text{'s velocity} : P\text{'s velocity}.$

3rd. In the system where each pulley hangs by a separate string, and the strings are parallel (Prop. XIII.), if W be raised through an inch, and P have also moved through such a space that the strings are kept tight, A will have been raised through one inch, and B through two inches (by the first case proved in this Proposition). And B having been raised through two inches, C (by the first case) will have moved through 2×2, or 2^2, inches; and the next moveable pulley (the *fourth*) will have been raised through 2×2^2, or 2^3, inches. By the same reasoning, if n were the number of *moveable* pulleys, the highest of them will have moved through 2^{n-1} inches, and the end therefore of the string by which P acts, and therefore P itself, through 2^n inches.

But $P : W :: 1 : 2^n$, by Prop. XIII., since they balance each other,

$$:: 1 \text{ inch} : 2^n \text{ inches,}$$

$$:: \text{space described by } W : \text{space described in same time by } P,$$

$\therefore P : W :: W\text{'s velocity} : P\text{'s velocity.}$

4th. In the case of the Inclined Plane, let the weight (W) be kept at rest at D on the inclined plane AB by the power (P) which acts parallel to the plane by means of a string DP; then, if P be made to move through the space Pp, W will move through an equal space DG along the plane. Through G draw GH horizontal, and through D draw 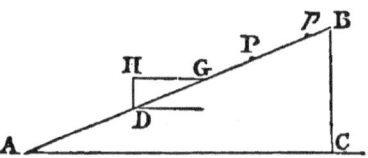 DH vertical. Then, W, by being moved through DG, has been raised vertically, that is, in the direction of gravity, through DH. Since, therefore, in the time that P moves, through a space equal to DG, W moves *vertically* through DH, DG is to DH as the velocity of P in the direction of its action is to the velocity of W *in the direction of gravity*.

Now, in the triangles GDH, ABC, since GH is parallel to AC, angle DGH = alternate angle BAC; and since DH, being vertical, is parallel to BC, angle GDH = alternate angle ABC; also angle DHG = a right angle = angle BCA, $\therefore$ the triangles are equiangular, and similar.

But $P : W :: BC : AB$, by Prop. XIV., since they balance each other,

$$:: DH : DG, \text{ by similar triangles,}$$

$\therefore P : W :: W$'s vel. in direction of gravity : P's vel. in the direction of its action.

Cor. In each of the above cases

$P \times P$'s velocity $= W \times W$'s velocity in direction of gravity.

Questions on Chap. IV.

(1) In the *Wheel and Axle*, what is the difference between *axle* and *axis*? Is the *axis* fixed, or rotatory?

(2) In the *Wheel and Axle*, is there any advantage in having the rope, which passes round the *Wheel*, thicker than that which passes round the *Axle*?

> (3) In the *Wheel and Axle*, the radius of the *Wheel* being three times that of the *Axle*, and the rope on the *Wheel* being only strong enough to support a tension of 36 lbs., what is the greatest weight which can be lifted?

(4) In Prop. x. how will the proof be affected if P and W, instead of acting vertically, act by means of strings in any other directions?

(5) What *Mechanical Powers* are employed in a *Crane* of the ordinary construction?

(6) In the single moveable Pulley, is any mechanical advantage gained, if the weight of the pulley be not less than the *power*?

(7) Why is it easier to move a heavy body when it is placed upon rollers, than to draw it along a rough horizontal plane?

(8) In Prop. xii., if the number of strings at the lower block be 6, what limit must be put to the weight of the lower block, so that any mechanical advantage may be gained by this system of pulleys?

(9) In Prop. xiii., if a weight of 1 lb. be supported by 1 oz., what is the number of moveable pulleys? Draw a figure to represent this case.

(10) In Prop. xiii., will the ratio $P : W$ be increased or diminished by taking into account the weights of the *strings*?

(11) In the proof of Prop. xiv., where is it assumed, that the *Inclined Plane* is perfectly *smooth*?

(12) Why is it easier to push a heavy body up a smooth *Inclined Plane* than to lift it through the same vertical height?

> (13) A smooth *Inclined Plane* rises $3\frac{1}{2}$ feet for every 5 feet of its length, what force must a man exert parallel to the plane, to prevent a weight of 200 lbs. from slipping down?

(14) A railway train travels over 150 miles in 5 h. 40 m., what is its average *Velocity* in feet per second?

(15) A race of 2 miles, 3 furlongs, and 62 yards, was run in 4 min. 12 sec. What was the average *Velocity* in feet per second?

(16) A body moves uniformly through 40 feet in 3 seconds, and another body through 25 yards in 6 minutes. What is the ratio of their *Velocities?*

(17) The earth's radius at the Equator is 3962·8 miles; and it makes a complete revolution about its axis in 23 h. 56 min.; what is the *Velocity* of a point at the Equator in feet per second?

(18) In Prop. xv., are P and W supposed at first to balance each other? If so, what puts the machine in motion? Is the motion *uniform?*

(19) In Prop. xvi., Case 3, if W be made to descend with a given velocity (v), what will be the *Velocities* of the several *pulleys?*

(20) In Prop. xvi., Case 3, would the result be true, if the pulleys were of different *sizes?*

CHAPTER V.

THE CENTRE OF GRAVITY.

49. *Definition of* CENTRE OF GRAVITY.

The *Centre of Gravity* of any body, or system of bodies, is that point upon which the body, or system, acted on only by the force of gravity, will balance itself *in all positions*.

This definition supposes all the particles of the body, or system of bodies, to be rigidly connected; and the point, called the *Centre of Gravity*, is also supposed to be in rigid connection with all the parts of the system, while that point is itself maintained *at rest*.

The *Centre of Gravity* of a body, or system, is, as it were, the *fulcrum*, round which the body, or system, when placed in any position, has, *of itself*, no tendency to turn, although the body be *capable* of being moved in any way about that fulcrum.

CENTRE OF GRAVITY. 39

50. PROP. XVII. *If a body balance itself upon a line in all positions, the Centre of Gravity of the body is in that line**.

Let the body AB balance itself *in all positions* upon the straight line CD, the *Centre of Gravity* of the body shall be in CD.

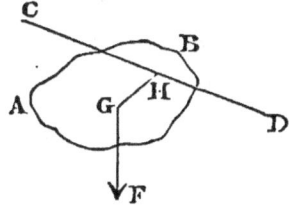

For, if not, let a point G, without CD, be the *Centre of Gravity;* and, first bringing CD into an *horizontal* position, turn the body round CD until G is in the same horizontal plane with CD; then draw GH perpendicular to CD, meeting it in H, and GF vertical.

Now, since, by Definition, the body will balance itself on G *in all positions*, it will balance itself on G in *this* position, that is, the resultant of all the forces acting on the body passes through G; and since these forces (being the pressures exerted upon the several particles of the body by the force of gravity) are all parallel and vertical, the resultant will also be vertical, and equal to the *sum* of them, viz. the weight of the body. Replacing, then, *all* the forces acting on the body by their *resultant*, we have the case of a single force acting perpendicularly at G to turn the lever GH round the fulcrum H (CD, and therefore H, being supposed fixed in position); which force is not counteracted by any other, and therefore will turn the body round H, that is, round CD. But, by supposition, the body *balances* itself upon CD *in all positions*. Hence, the assumption that G lies anywhere without CD leads to an impossibility; and therefore G can only be *in CD*.

51. PROP. XVIII. *To find the Centre of Gravity of two heavy points*†, *and to shew, that the pressure at the*

---

* That is to say—If there be a line round which, as an axis, a body can be made to revolve, so that, when the line is held in *any* position, the body, after being made to revolve round it into *any* position, remains at rest, the *Centre of Gravity* of the body is in that line.—It is evident, that the line must *pass through* the body.

† By "heavy points", in this Proposition and the next, are meant exceedingly small material bodies, and not *geometrical* points. For a *geome-*

Centre of Gravity is equal to the sum of the weights in all positions.

Let P and Q be the weights of two heavy points A and B, supposed to be connected by a straight rigid rod AB without weight.

In AB take a point C, such that $BC : AB :: P : P+Q$; then, *dividendo*,

$BC : AB-BC :: P : \overline{P+Q}-P$,

or $BC : AC :: P : Q$.

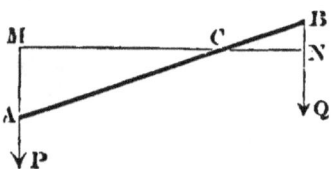

Through C draw MCN horizontal; and through A and B draw the vertical lines PAM and QNB; these last are the lines in which the *weights* P and Q act.

Then, the angles at M and N being right angles, and the angle ACM being equal to the opposite angle BCN, the angle CAM is equal to the angle CBN, and the triangles ACM and BCN are equiangular, and ∴ similar.

Now, $P : Q :: BC : AC$,

:: $CN : CM$, by similar triangles;

therefore, if ABC be considered as a *lever* with fulcrum C, since P and Q are inversely as the perpendiculars drawn from the fulcrum to the lines in which the forces act, by the converse of Prop. VI. (Art. 26), P and Q will balance each other on C.

Also, if AB be turned round C into *any* other position, the same reasoning holds; and therefore A and B will balance on C in *all* positions of AB. Hence, by Definition, C is the *Centre of Gravity* of A and B.

Again, by what has been shewn, the weights P and Q will balance on C, when ACB is *horizontal*. But, in that case, by Axiom II. (Art. 19), the pressure on the fulcrum is equal to *the sum of the weights*. And in *any* other position of AB, as in the fig., P and Q will produce

---

trical point does not possess length, or breadth, or thickness, and consequently can be of no *weight*, since it can contain no matter.

the same effect as if they acted at M and N on the horizontal straight lever MCN, exerting a pressure on C equal to their sum. Therefore the pressure at the *Centre of Gravity* is equal to the sum of the weights *in all positions* of the system.

52. Prop. XIX. *To find the Centre of Gravity of any number of heavy points; and to shew, that the pressure at the Centre of Gravity is equal to the sum of the weights in all positions.*

Let A, B, C, three heavy points whose weights are P, Q, R, be connected together and placed in any position.

Join AB, and take D a point in AB, such that

$$BD : AB :: P : P+Q;$$

$\therefore BD : AB - BD$, or AD, $:: P : \overline{P+Q} - P$, or Q;

therefore, by Prop. XVIII., D is the *Centre of Gravity* of A and B; and the pressure produced by P and Q, in all positions of the system, is a pressure $P + Q$ acting vertically at D.

Join DC, and in DC take a point E, such that

$$DE : DC :: R : P + Q + R;$$

$\therefore DE : EC :: R : P + Q;$

therefore E is the *Centre of Gravity* of the weight $\overline{P+Q}$ acting at D, and R acting at C; and if E be supported, those weights are supported in any position of the system. Since therefore the system will balance itself in all positions on E, that point is its *Centre of Gravity*;—and the Pressure on E is $P + Q + R$.

The construction here applied to a system of *three* bodies may be extended to a system of *any number* of bodies.

Wherefore the Centre of *Gravity of any number* of heavy points may always thus be found, and the pressure on the *Centre of Gravity* is equal to the sum of the weights in all positions.

53. By the definition given in Art. 49 of "*the Centre of Gravity*" of a body, it will be understood, that to have a *Centre of Gravity* a body must have *Weight*. Now in the next two Propositions it is required to find the *Centres of Gravity* of a *line*, and of a *plane*, the former of which is defined by Euclid to have *length* merely, without either *breadth* or *thickness;* and the latter, though possessing *length* and *breadth*, is defined to be without *thickness*. A geometrical line, or plane, therefore, *can* have no weight; since there can be no weight where *matter* does not exist, and when *matter* exists under any form whatever, it is of *three* dimensions, or has *length, breadth*, and *thickness*. The line, therefore, and the plane, of which it is required to find the *Centres of Gravity*, are not the line and plane of Geometry.

But the line of which the *Centre of Gravity* is determined in the next Proposition is supposed to be formed of very small *equal* heavy bodies placed either in contact, or at equal distances, along the whole length of the line. And the plane triangle referred to in the next Proposition but one, is supposed to be made up of such lines, arranged parallel to any one of the sides of the triangle, and at equal distances from one another.

54. PROP. XX. *To find the Centre of Gravity of a straight line.*

Let AB be a straight line composed of small equal heavy bodies ranged either in contact, or at equal distances, along its whole length.

Bisect AB in C, and let P and Q be two of the small heavy bodies equally distant from C. Then, by Axiom I. Art. 19, and Prop. VII., P and Q will balance in every position on C. And since the same is true of all such pairs of heavy bodies that are equidistant from C, the whole line will balance on C in every position, and therefore C is the centre of gravity of the line.

55. PROP. XXI. *To find the Centre of Gravity of a triangle.*

Let ABC be a triangle formed of lines ranged, at equal

distances, parallel to any one of the sides, the lines themselves being made up of small equal heavy bodies placed in contact, or at equal distances. Let bfc, parallel to BFC, be one of these lines.

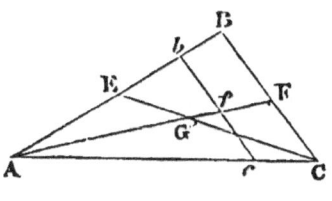

Bisect AB in E, and BC in F; join AF, CE by lines intersecting in G;—G is the *Centre of Gravity* of the triangle.

For $Af : fb :: AF : FB$ (from the equiangular triangles Abf, ABF),

$\qquad :: AF : FC$, $\because BF = FC$;

$\qquad :: Af : fc$ (from the equiangular triangles AFC, Afc).

And since the first and third terms of this proportion are the same, fb is equal to fc; and therefore the straight line bc would balance in any position on f, by Prop. XX.

In the same manner all the other lines parallel to BC may be shewn to balance in any position on the points in which they are cut by AF; therefore the whole triangle will balance on AF in any position. Hence the *Centre of Gravity* of the triangle is in AF, by Prop. XVII.

Similarly, by supposing the triangle to be made up of lines parallel to AB, it may be shewn, that the *Centre of Gravity* of the triangle is in the line CE.

But AF and CE have only one point in common, that is, G, the intersection of AF and CE: therefore G is the *Centre of Gravity* of the triangle ABC.

56. PROP. XXII. *When a body is placed on a horizontal plane, it will stand or fall, according as the vertical line, drawn from its* Centre of Gravity, *falls within or without its base.*

[DEF. By "base" is here meant the area formed by drawing a string tightly round the lowest part of the body *in contact with the plane*. Thus, if the body be a three-legged stool, its '*base*', for this purpose, will be a *triangle*—of a chair the '*base*' will be a *quadrilateral;* and so on.—Again, the '*base*' of a man, standing on a horizontal plane, will be greater or smaller, according as his feet are more or less apart, although the portion of him in contact with the plane remains the same.]

Let $ABCDE$ be the '*base*' of the body (according to Definition), G the Centre of Gravity of the body, GH a vertical line through G meeting the horizontal plane, on which the body is placed, in H.

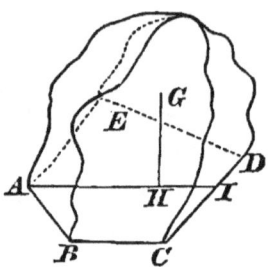

Join AH, and produced, if necessary, let it meet the boundary, CD, of the 'base' in I.

1st. Let H fall *within* the 'base'; it is therefore *between* A and I. Then, since the whole weight (W) of the body may be supposed concentrated at G, and acting in the direction GH, AHI being regarded as a horizontal *Lever*, the only tendency of W, acting at H, would be to produce motion *downwards* round A, which motion the resistance of the plane will prevent. For the same reason no motion of the *Lever* can take place round I. Therefore the body will fall over neither round A nor I.

In the same manner it may be shewn, that the body will not fall over round either extremity of *any other* horizontal straight line drawn through H, and terminated by the boundary of the 'base'. Hence the body will *stand*.

2nd. Let H fall *without* the 'base'; then, regarding AIH as a Lever, no motion can take place round A as fulcrum, because the tendency of W, acting at H is to draw AH *downwards*, a motion which the plane prevents. But if I be supposed the fulcrum, W acting at H will cause the arm AI to move *upwards* from the plane round I, and as there is no counteracting force, the body will *fall* over round I.

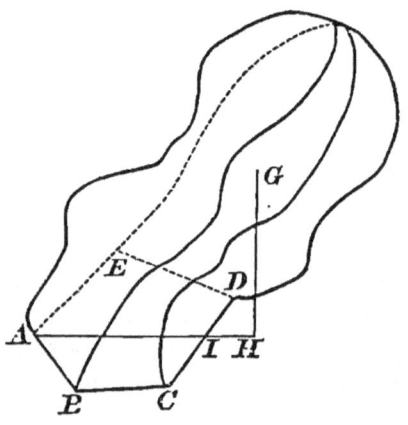

CENTRE OF GRAVITY. 45

57. Prop. XXIII. *When a body is suspended from a point* it will rest with its Centre of Gravity in the vertical line passing through the point of suspension.*

Let AB be the body in *any* position, C the point from which it is suspended, G its *Centre of Gravity*, NGK a vertical line drawn through G, CN a perpendicular from C on NGK.

Since the weight of the body acts in the vertical line which passes through G, the *Centre of Gravity*, it may be supposed to be applied at any point in that line, and therefore at N; and, CN being considered as a *Lever* moveable round a fulcrum C, as there is no force to counteract W acting perpendicularly at N on CN, motion must ensue.

But if NG pass through C, CN vanishes, the weight is sustained by the immoveable fulcrum C, and the body is *at rest*, that is, when G is in the vertical line passing through C.

Cor. This Proposition points to an easy method of finding the *Centre of Gravity* of a body, which may be applicable in certain cases.

For if the body, whose *Centre of Gravity* is required, be suspended from a point in itself, and brought to a state of rest, and a vertical line can be drawn on its surface through the point of suspension, the *Centre of Gravity* of the body is somewhere in that line. If then the body is suspended from some *other* point in itself, and the vertical line, as before, be drawn through this point, the *Centre of Gravity* is also in this line. Hence it follows, that the intersection of the two lines so drawn on the body is the *Centre of Gravity* required.

Questions on Chap. V.

(1) Can a body, or system of bodies, have more than one '*Centre of Gravity*'? Must there *always* be *one?*

(2) Can the *Centre of Gravity* of any body be *outside* of it? If so, what is the precise meaning of the term in that case?

---

* That is, about which point it can swing freely.

(3) Has an empty bottle a *Centre of Gravity?* How does the definition of *Centre of Gravity* apply to such a case?

(4) In a system of several unconnected bodies, how can you speak of *their Centre of Gravity?*

(5) In Prop. XVII., is the line *CD* within, or without, the body?

(6) In Prop. XVIII., how do the *points* differ from *points* as defined by Euclid?

(7) In Prop. XIX., the *points* being *heavy* points, are the *lines* connecting them *heavy* lines? What do you mean by a *heavy line?*

(8) In Prop. XX., what is the *straight line* made of?

(9) In Prop. XXI., is it the *Centre of Gravity* of the three *sides* of the triangle, or of the *area* bounded by those sides, which is found? Would the *Centre of Gravity* in the one case *differ* from that in the other?

(10) In Prop. XXI., the triangle is supposed to have no thickness; how would the *Centre of Gravity* be found, if the triangle were cut from a board of half an inch uniform thickness?

(11) Can you deduce the *Centre of Gravity* of a parallelogram from that of a triangle? What would the result be *in all cases?* and the practical rule?

(12) In Prop. XXII., what is the exact meaning of '*base*'? What is the '*base*' of a horse? Can the same horse make its '*base*' greater or smaller at pleasure, within certain limits?

(13) If a wall be well built, but out of plumb, under what conditions will it stand?

(14) Why do we, in ascending a hill, appear to lean forwards; and in descending, to lean backwards?

(15) Of what use is the long pole to a rope-dancer?

(16) How would you find the *Centre of Gravity* of a thin *slate* of irregular outline?

(17) Can you *shew why*, of carriages which have the same '*base*', that which has its *Centre of Gravity* the highest is the most easily overturned?

(18) Does the position of the *Centre of Gravity* of a body depend upon the *form* of the body, or upon its *weight?*

HYDROSTATICS.

[*The explanatory matter, printed in small type, forms no actual part of the* UNIVERSITY COURSE; *but is illustrative of the particular Definition, or Proposition, which it immediately follows; and will be found useful for answering the Questions, and solving the Problems, which are usually given in the University Examinations.*]

CHAPTER I.

58. HYDROSTATICS is the Science which treats of the effects of pressures applied to, or produced by, *Fluids in a state of rest.*

59. *Definitions of* FLUID; *of* ELASTIC *and* NON-ELASTIC FLUIDS.

A FLUID is a material body which can be divided in any direction, and the parts of which can be moved among one another by any force, however small.

Mercury, Water, Air, Gas, Steam, are all instances of *Fluids.*

FLUIDS have been divided into *Elastic,* and *Non-elastic.*

(1) ELASTIC FLUIDS are those of which the dimensions are increased or diminished when the pressure upon them is diminished or increased.

Air is an *Elastic* Fluid, as will hereafter be proved. So also is Gas, and Steam.

(2) NON-ELASTIC FLUIDS are those of which the dimensions are not diminished by the addition of pressure, or increased by the withdrawal of pressure.

48   HYDROSTATICS.

Water, *Mercury*, and probably all other *liquids*, are *compressible*, but in a slight degree only. They offer, however, so much resistance to compression, that the conclusions obtained on the supposition of their being entirely *incompressible* are free from any sensible error, except in cases where the pressure is exceedingly great.

CHAPTER II.

PRESSURE OF NON-ELASTIC FLUIDS.

60. PROP. I. *Fluids press equally in all directions.*

Let a close vessel of any shape be filled with fluid, and let A, B, C, and D, equal portions (of any size and shape) of the sides of the vessel, be removed, and their places supplied by pistons* fitting the orifices exactly, and acted upon by pressures just sufficient to keep them at rest. Then, if an additional force P be applied to any one of the pistons, it is found, by experiment, that an *equal* additional pressure must be applied to *every other* of the pistons, to prevent the fluid from bursting out.

This proves that a pressure communicated to the surface of a fluid at rest is transmitted by the fluid undiminished to every other equal portion of the surface by which the fluid is bounded.

Also, whatever be the *directions* in which any of these pistons are inserted through the sides of the vessel, the same pressure is required to keep each of them at rest; which shews that, at the same point in a fluid, the fluid presses with *equal* force in *all* directions.

61. PROP. II. *The pressure upon any particle of a fluid of uniform density is proportional to its depth below the surface of the fluid.*

---

* A PISTON is a solid plug *exactly* fitting an opening into a vessel filled with fluid, not tight like the bung of a cask, but capable of moving *freely*.

PRESSURE OF NON-ELASTIC FLUIDS.

Let the vertical lines *PM* and *QN* be drawn through *P* and *Q*, two equal particles situated in the interior of an uniformly dense fluid which is at rest; and let them meet the surface of the fluid in the points *M* and *N*.

Suppose the pressures on *P* and *Q* to be produced by the *weight* of fluid particles, of the same magnitude as *P* and *Q*, placed in contact with each other along the lines *PM* and *QN*. Then, since the fluid is of uniform density, the weights of such lengths of equal particles will be proportional to their number of particles, that is, to the *lengths* of the lines.

Hence, pressure on the particle *P*

: pressure on the particle *Q*

:: weight of the line *PM* of particles

: weight of the line *QN* of particles,

:: *PM* : *QN*.

Again, if a *piston* be inserted through the side of an open vessel containing fluid, so as to test the amount of the pressure at any point *R*, it is found, by experiment, that whether a line of fluid particles extends vertically to the surface, or a portion of such line is cut off by the side of the vessel intervening, the pressure at *R* is the same.

Hence, in either case, the pressure at any point is proportional to the *vertical* depth of the point below the surface.

62. PROP. III. *The surface of every fluid at rest is horizontal.*

[The manner in which the matter composing *Non-elastic Fluids* acts is two-fold; first, by the particles pressing with their *weights* on those immediately below them; and, second, by the property these particles possess of communicating, *in all directions,* and without diminution, any pressure to which they are subjected.]

Let there be two contiguous particles, the centres of which are the points P and Q, situated in the same horizontal plane in the interior of a fluid at rest. Since the action of gravity on these particles is at right angles to the horizontal plane in which they lie, its action on either of them can produce *no* effect *in* that plane. It is in consequence, therefore, of the action of the fluid which *surrounds* them, that the particles have no horizontal motion: and this action must be the *same* on each, otherwise motion would ensue.

But since fluids press equally in *all* directions (Prop. I.), and the *horizontal* pressures on the two particles are *equal*, the *vertical* pressures on them, which are as the distances Pa and Qb below the surface of the fluid (Prop. II.), must also be equal; and therefore

$$Pa = Qb;$$

∴ ab is parallel to PQ, and is ∴ *horizontal*.

And since, in like manner, the line joining *any* two adjacent particles in the surface of the fluid may be shewn to be *horizontal*, the *surface* itself is horizontal.

Cor. Hence also it appears, that "*In a fluid at rest, acted on only by the force of gravity, the pressure is the same at every point in the same horizontal plane*".

63. Prop. IV. *If a vessel, the bottom of which is horizontal and the sides vertical, be filled with fluid, the pressure upon the bottom will be equal to the weight of the fluid.*

The sides of the vessel being vertical, and the whole of the fluid being conceived to be made up of vertical straight lines of fluid particles, each of these lines will press vertically with its *weight;* and the *sum* of these vertical pressures will be the *weight* of the whole fluid in the vessel. Now the base of the vessel, being horizontal, will counterbalance all the vertical pressures upon it, and destroy them entirely. The pressure, therefore, sustained by the horizontal base of a vessel, whose sides are vertical, is equal to the weight of the fluid in the vessel.

PRESSURE OF NON-ELASTIC FLUIDS.

64. From Propositions II. and IV. the following conclusion may be drawn:—

The pressure of a fluid, on any horizontal plane placed in it, is equal to the weight of a column of the fluid whose base is the area of the plane, and whose height is the depth of the plane below the horizontal surface of the fluid.

Let AB and CD be two equal areas in the same horizontal plane immersed in a fluid. Since the pressure is the same on every point in the same horizontal plane, and (the areas AB and CD being equal) the number of particles in the plane CD is the same as that in the plane AB, therefore the whole pressure on AB = whole pressure on CD.

Let AB be such an area, that a vertical line of fluid particles reaches from *every* particle in it to the surface; and suppose the rest of the fluid to become solid; $EABF$ then may be considered as a *vessel* with vertical sides and horizontal base, and the pressure on AB = weight of the column EB of fluid, by Prop. IV.

Therefore, the pressure on *any horizontal* area CD = weight of a column of the fluid whose base is CD, and whose height is the vertical depth of CD below the horizontal plane of the surface of the fluid in the vessel.

65. From Art. 52 it appears, that the pressure on the *Centre of Gravity* of a system of bodies, considered as heavy points, is the same as if the weights of the bodies were collected at it. So long, therefore, as the *Quantity of Matter* in the system remains the same, the amount of *Pressure* produced in the direction of gravity by the weights of the several parts of the system is the same, whatever be the manner in which they may be arranged.

Now though, at first sight, it might appear probable that the pressure on the bottom or the sides of a vessel, filled with fluid, would depend (somehow or other) on the *quantity* of the fluid by which that pressure is produced, such is not the case; for it is found, both from experience and by theory, that the pressure produced by a fluid, at rest, on the inner surface of the vessel containing it, is not dependent on the *quantity* of the fluid—a fact apparently so much at variance with the law governing the amount of pressure produced by *solid* bodies, that it has been called the HYDROSTATIC PARADOX.

4—2

66. Prop. V. *To explain the Hydrostatic Paradox.*

The Hydrostatic Paradox is this:—

"*Any pressure, however small, may be made to counter-balance any other Pressure, however great, by means of a small quantity of fluid*".

The top and bottom of a vessel AD are boards which are connected together by leathern sides. The vessel communicates with a vertical tube EF of small uniform bore, by means of a horizontal pipe CE. Let A be the number of square inches in the area of the board AB, and a the number of square inches in the area of a horizontal section of the pipe EF.

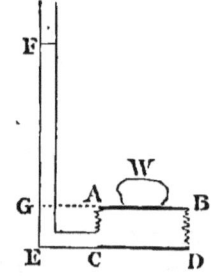

Let AB be held in a horizontal position, and water be poured into the tube until it just rises in AD to AB. The water will therefore rise in the tube to G, a point in the same horizontal plane as AB. Prop. III.

If now there be a heavy weight of W lbs. laid upon AB, it is found, that it can be supported at rest by a small additional column FG, of water poured into the tube.

To shew the reason of this;

Since the sides of the tube FG are vertical, the pressure on the horizontal section of the tube at G is the weight of the fluid column FG. Suppose w to be this weight, in pounds; then, since the pressure at every point in the same horizontal plane of a fluid at rest is the same, and assuming the fluid to be made up of small equal particles,

Pressure on the area A of AB : pressure on the area a at G :: N°. of particles in A : N°. in a :: $A : a$;

or $W : w :: A : a$;

$$\therefore w = W \times \frac{a}{A}.$$

If, therefore, W be given, and however great it may be, w may be made as small as we please by diminishing a, or increasing A; that is, by adjusting the dimensions of A and a, *any pressure however small may be made to balance any other pressure however great.*

Whether the pressure on the area a at G be produced by the weight of the fluid column GF, or by means of a piston acted on by some force, the pressures on the areas, a at G, and A at AB, will still bear to one another the ratio of a to A, a ratio which is wholly independent of the quantity of fluid contained in the vessel.

67. PROP. VI. *If a body floats on a fluid, it displaces as much of the fluid as is equal in weight to the weight of the body; and it presses downwards, and is pressed upwards, with a force equal to the weight of the fluid displaced.*

Let $ABCD$ be a body *at rest* that displaces the portion $BEDCB$ of the fluid on which it *floats*.

The *weight* of the floating body produces a pressure which acts vertically downwards. Therefore the pressure of the fluid which keeps the body at rest must act vertically upwards, and be equal to the *weight* it balances.

Now suppose the floating body to be removed, and the space $BEDCB$ filled with fluid of the same kind as the surrounding fluid; the *equilibrium* of the fluid will not be disturbed; neither will the pressure of that part of it which was formerly in contact with the surface of the floating body be altered, if the particles of fluid in $BEDCB$ be supposed to become permanently connected with one another, and to form a solid.

Let this take place; then the pressure downwards of the part $BEDCB$ of the fluid which becomes solid is its *weight*. And since this pressure is counteracted by the same sustaining power as that which balanced the weight of the floating body, *the weight of the floating body must be equal to the weight of the fluid it displaces.*

QUESTIONS ON CHAPTERS I. AND II.

(1) What are the characteristic differences between a *Fluid* and a *Solid?*

(2) If a *Fluid* be a '*material body*', according to the Definition, how can *Steam* be a *Fluid?*

(3) If a 'material body' be compressible, is it necessarily *elastic?*

(4) Is *Water elastic,* or *inelastic?*

(5) If "*Fluids press equally in all directions*", does this mean, that a Fluid, acted on only by the Force of Gravity, will press *upwards* as well as *downwards?*

(6) In Prop. II. the pressure of the Atmosphere on the surface of the fluid is not taken into account; is the truth of the proposition affected thereby?

(7) *Pressure* is always the result of *force;* what then is the *force* supposed to be acting in Prop. II.?

(8) If the pressure be different at different points of a fluid at rest, how can it be of '*uniform density*'?

(9) Is it assumed in Prop. III. that the fluid is of *uniform density?* If so, where?

(10) What *forces* are supposed to act on the fluid in Prop. III.?

(11) What is meant by the '*surface*' of a fluid in Props. II., and III.?

(12) In Prop. IV., the pressure of the *atmosphere* on the surface of the fluid is not reckoned; will not that greatly affect the pressure on the bottom of the vessel?

(13) How can the pressure on the bottom of a vessel filled with fluid, and acted on only by the force of gravity, be *greater than* the whole weight of the fluid?

(14) *Bramah's press,* used by packers and others, is constructed on the principle explained in Prop. V.; what are the practical limits to its power?

(15) Why is it necessary, that the fluid used in the *Hydrostatic Paradox,* and in *Bramah's press,* should be *non-elastic?*

(16) If the floating body in Prop. VI. were *wholly* immersed, and at rest when left to itself, would Prop. VI. hold true?

(17) If a vessel be quite filled with fluid, and a solid body be put into it, which floats, and "*displaces as much of the fluid as is equal in weight to the weight of itself*", will the solid body increase the pressure on the bottom of the vessel, or not?

CHAPTER III.

SPECIFIC GRAVITIES.

68. THE BULK, or VOLUME, or CONTENT, or MAGNITUDE, of any body is measured by the number of times it contains that of some other body previously fixed upon as a standard of magnitude, or *unit*.

A cube, whose edge is an inch in length, is called "a cubic inch". In the following pages *a cubic inch* will be taken for the *unit* of solid measurement; so that when it is said, that M is the *bulk, volume, content,* or *magnitude,* of a body, it is meant that the number of cubic inches in the body is the number of units in M.

69. *Definition of* SPECIFIC GRAVITY.

The SPECIFIC GRAVITY of any substance is the weight of a *unit* of its magnitude, or volume.

If, as stated in the last Article, the magnitude of a body be measured by the number of *cubic inches* it contains, and the weight of one cubic inch be given in *grains,* then it will follow, that, S being taken to represent the *Specific Gravity* of any substance, S is the *number* of *grains* that *one cubic inch* of that substance *weighs.*

The Tables, which are called "TABLES OF SPECIFIC GRAVITIES", give the *Ratios* which the weights of bulks of various substances bear to *equal* bulks of *water*. In other words, they give the number of times that the weight of any bulk of each of the substances contains the weight of an *equal* bulk of *water*.

It having been found by experiment that the weight of a piece of Iron : the weight of a bulk of Water *of the same size* :: 7·8 : 1, and that the weight of a piece of Silver : weight of an equal bulk of Water :: 10·5 : 1, and so for other substances, *Tables* have been formed, in which the numbers 1, 7·8, 10·5, &c., are placed opposite the words, "Water", "Iron", "Silver", &c. By means of these *Tables* (as will be shewn), the *weight* of any bulk of any of the substances so registered can be determined, if the weight be known of some particular bulk of any one of them.

. The numbers given in these *Tables* are generally called the "*Specific Gravities*" of the several substances registered; but the enunciation of Prop. VII. Art. 70 will not permit them to be so called here.

The "TABLES OF SPECIFIC GRAVITIES" give—

Platinum	21·53	Diamond	3·52
Gold	19·4	Sea Water	1·027
Mercury	13·6	Water	1·
Lead	11·4	Proof Spirit	0·93
Silver	10·5	Pure Alcohol	0·825
Copper	8·9	Air, at the surface of the Earth—the average	0·00125
Iron	7·8		
Tin	7·3		
Zinc	6·9	Ice	0·926

Hence it may be shewn, that—

i. *The weights of any two substances, are of equal bulk which are in the ratio of the numbers given by the Tables as corresponding to the substances.*

For, w, w', w'', being the respective weights of equal bulks of Water and of any two substances, as Iron and Silver, since, as explained above,

$$w' : w :: 7\cdot8 : 1, \text{ and } w : w'' :: 1 : 10\cdot5;$$

therefore, compounding these proportions,

$$w' : w'' :: 7\cdot8 : 10\cdot5.$$

So that, if it be required, for example,

ii. *To find the weight of a cubic foot of Iron, having given that the weight of 10 cubic inches of Silver is 61 ounces nearly—*

∵ Weight of a cubic foot (or 12×12×12 cubic inches) of Silver

$$= 12\times12\times12\times\frac{61}{10} \text{ ounces};$$

∴ Weight of a cubic foot of Iron $= \dfrac{12\times12\times12\times61}{10} \times \dfrac{7\cdot8}{10\cdot5}$ oz.

$$= \frac{12\times12\times12\times61\times7\cdot8}{105} \text{ oz.} = 7830 \text{ ounces, nearly.}$$

70. PROP. VII. *If M be the Magnitude of a body, S its Specific Gravity, and W its Weight, $W = MS$.*

Suppose the *unit* of the measurement of magnitude to be a *cubic inch*; then

$M =$ *number* of cubic inches in the body.

And the Specific Gravity (S) is the weight of *one* cubic inch;

$\therefore$ the whole weight of the body $= M \times S$,

or, $W = MS$.

71. *To find the relation which exists between the Weights, Magnitudes, and Specific Gravities, of two substances and of a compound formed of them.*

Let $W, M, S, \ W', M', S', \ W'', M'', S''$, be the Weight, Magnitude, and Specific Gravity, of each of the two substances, and of the compound, respectively.

Then, it being supposed that the portions of the substances which are combined together lose neither bulk nor weight by being mixed,

i. $M'' = M + M'$,

ii. $W'' = W + W'$;

$\therefore$ by Art. 70, iii. $M''S''$, or $(M+M')S'' = MS + M'S'$.

And it might be shewn, that similar relations exist between the Weights, Magnitudes, and Specific Gravities, of the several substances and the compound formed of them, whatever be the *number* of the simple substances.

72. Cor. Let $\sigma, \sigma', \sigma''$, be the numbers which are attached to the names of the two simple substances and the compound, respectively, in the "Tables of Specific Gravities", the S. G. (specific gravity) of water being 1.

By Art. 71, $(M+M')S'' = MS + M'S'$;

$\therefore (M+M') \cdot \dfrac{S''}{S} = M + M' \cdot \dfrac{S'}{S}$.

But $S' : S :: $ weight of a cubic inch of the second substance

: weight of a cubic inch of the first substance,

$:: \sigma' : \sigma$; by Art. 69, i.

Similarly $S'' : S :: \sigma'' : \sigma$,

$\therefore (M+M') \cdot \dfrac{\sigma''}{\sigma} = M + M' \cdot \dfrac{\sigma'}{\sigma}$,

or $(M+M')\sigma'' = M\sigma + M'\sigma'$.

Cor. In like manner, if W, W', W'', be the weights, in pounds, of the two substances and their compound, respectively, it may be proved, that

$$\frac{W''}{\sigma''} = \frac{W}{\sigma} + \frac{W'}{\sigma'}.$$

73. Prop. VIII. *When a body of uniform density floats on a fluid, the part immersed : the whole body :: the specific gravity of the body : the specific gravity of the fluid.*

Let a body of uniform density float on a fluid with M cubic inches of it above, and N cubic inches below, the horizontal plane of the fluid's surface.

Let $S =$ S. G. (Specific Gravity) of the solid; $S' =$ S. G. of the fluid.

Then, $(M+N) \times S =$ weight of the solid by Prop. VII.,

$N \times S' =$ .... fluid displaced.

But because the body floats, these two weights are equal, by Prop. VI.;

$$\therefore NS' = (M+N)S; \text{ and } \therefore \frac{N}{M+N} = \frac{S}{S'},$$

or $N : M+N :: S : S'$;

that is, the part immersed : the whole body
:: S. G. of the body : S. G. of the fluid.

74. Prop. IX. *When a body is immersed in a fluid, the weight lost : whole weight of the body :: the specific gravity of the fluid : the specific gravity of the body.*

Let M be the number of cubic inches in a body of uniform density, which is wholly immersed in a fluid; S the S. G. of the body, S' the S. G. of the fluid.

Then the pressure downwards of the solid is its *weight;* and if the solid be removed, and the space it filled be occupied by an equal bulk of the fluid, equilibrium will still exist. And if the fluid so added become solid, the equilibrium will continue, and the pressure upwards of the surrounding fluid will remain the same as before.

Now the pressure downwards of the portion of the *fluid* which becomes solid is its *weight*. And as the pressure *upwards* of the surrounding fluid supports this *weight*, that pressure must be exactly equal and opposite to it.

The pressure downwards, therefore, of the original solid, before it was immersed (i. e. its weight MS, Prop. VII.), must, by immersion of the solid, have been diminished by a pressure upwards, arising from the surrounding fluid, exactly equal to the weight of the fluid displaced,—which weight, by Prop. VII., is equal to MS'.

∴ Weight lost by the body : the whole weight of the body :: MS' : MS :: S' : S :: S. G. of the fluid : S. G. of the body.

75. It appears from the proof of the last Proposition, that the pressure of a fluid on a body wholly immersed in it acts vertically upwards, and is equal to the weight of the fluid which the body displaces.

If this pressure be less than the weight of the body—that is, if the S. G. of the fluid be less than that of the solid—the pressure downwards arising from the weight of the solid will be greater than the pressure of the surrounding fluid upwards, and the body will therefore *sink* to the bottom of the vessel.

But if the pressure of the fluid upwards be greater than the weight of the body immersed,—that is, if the S. G. of the fluid be greater than that of the solid,—the pressure upwards will be greater than the pressure downwards, and the body will therefore *rise*, until the conditions of Proposition VIII. are fulfilled, and will then *float*.

76. PROP. X. *To describe the Hydrostatic Balance; and to shew how to find the Specific Gravity of a body by means of it*,—1st, *when its Specific Gravity is greater than that of the fluid in which it is weighed*,—2ndly, *when it is* less.

The *Hydrostatic Balance* is the Common Balance, with a hook attached to the under part of one of its scales, so that bodies may be weighed either by putting them into the scale, or by suspending them from it and letting them be immersed in a fluid as here represented.

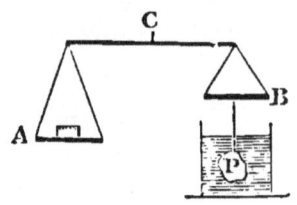

1st. Let the S. G. (Specific Gravity) of the body be *greater* than that of the fluid.

Since the S. G. of the solid is greater than that of the fluid, the body will sink in the fluid. Art. 75.

Let S = S. G. of the solid, S' = S. G. of the fluid.

W = weight required to balance the body when placed *in* the scale.

W' = weight required when the body is immersed;

∴ $W - W'$ = weight lost by immersion in the fluid.

Therefore, by Prop. IX.,

$$W : W - W' :: S : S';$$

$$\therefore S = \frac{W}{W - W'} \cdot S'.$$

But W and W' are known weights, and S' is supposed to be given, therefore S may be determined.

2ndly. Let the S. G. of the body be *less* than the S. G. of the fluid.

When the body, in this case, is *forced* under the surface of the fluid, the pressure downwards on it (which is the weight W of the body) being less than the pressure upwards (which is the weight of a quantity of fluid equal in magnitude to the fluid displaced), the body must on the whole be acted on by a pressure upwards equal to the difference between the weight of the fluid displaced and the weight of the solid.

SPECIFIC GRAVITIES.

To determine this last pressure, take a body Q,—called the *Sinker*,—of greater S. G. than that of the fluid, and large enough to sink both itself and the body P, whose S. G. is required, when P is attached to it. Let Q be first immersed by itself in the fluid, and balanced by weights in A. Next, leaving the weights in A, let P be attached to Q and both of them be immersed. The scale A will now preponderate, and to restore the equilibrium let a weight W'' be placed in the scale B*. This weight therefore is equal to the tendency *upwards* of P when immersed; that is,

W'' = weight of a bulk of the fluid equal to that of P *minus* the weight of P (W);

∴ weight of a bulk of fluid of the same size as P

$$= W + W''.$$

Now since, (by Prop. VII.), $W = M \times S$, when the magnitudes of the bodies are the same, $W \propto S$;

∴ weight of P : weight of an equal bulk of fluid

:: S. G. of solid : S. G. of fluid,

or, $\quad W : W + W'' :: S : S'$;

$$\therefore S = \frac{W}{W + W''} \cdot S'.$$

But W and W'' are known weights, and S' is supposed to be given; therefore S may be found.

It will be observed that it is not requisite for the exact weight of the *Sinker* to be known.

77. Prop. XI. *To describe the Common Hydrometer; and to shew how to compare the Specific Gravities of two fluids by means of it.*

---

* The better way in practice is to restore the equilibrium by removing a weight W'' from the scale A; but the method described in the text perhaps renders the demonstration easier to be understood.

The common *Hydrometer* consists of two hollow spheres attached to each other, and of a cylindrical slender stem, whose axis, if produced, would pass through the centres of both the spheres. The upper sphere is empty; and the lower is filled with lead or mercury, so as to make the instrument float steadily in a vertical position when put into a fluid. The stem is graduated by divisions of equal length.

The *Hydrometer* is made lighter than an equal bulk of any of the fluids whose *Specific Gravities* it is employed to compare.

Suppose the bulk of the portion of the stem included between every two graduations to be one four-thousandth part of the bulk of the whole instrument. When the *Hydrometer* floats vertically in a fluid whose S. G. is S, suppose 20 divisions are above the surface; and when it floats in a fluid whose S. G. is S', let there be 30 divisions out.

Now, the *weights* of the bulks which are displaced of the two fluids are the same, each being equal to the weight of the instrument, by Prop. VI. If M and M', therefore, be the magnitudes of the fluids displaced, by Prop. VII.,

$M \times S =$ weight of the *Hydrometer* $= M' \times S'$;

$\therefore S : S' :: M' : M,$

$:: 4000 - 30 : 4000 - 20,$

$:: 3970 \quad\quad : 3980;$

and the ratio of the *Specific Gravities* of the two fluids is thus determined.

78. A mark P is made at the point in the stem to which the instrument sinks in a fluid called "*Proof Spirit*", which is a mixture consisting of equal *weights*,—(*not* equal *magnitudes*),—of pure Alcohol and of Water. Alcohol being lighter than Water, if a mixture of these two fluids contain a greater weight of the former than it does of the latter, it will be lighter than an equal bulk of *Proof Spirit*, and the *Hydrometer* therefore will displace a greater bulk of it than it does of *Proof Spirit*, that is, it will sink deeper in

the mixture than in *Proof Spirit*. Wherefore the surface of such a mixture will rise to a higher point in the stem than P. In such a case the mixture is said to be "*above proof*".

But if the weight of the Water contained in the mixture be greater than that of the pure Alcohol, the *Hydrometer* will not sink so low as to the point P, and the fluid is then said to be "*below proof*".

Questions on Chap. III.

(1) When M is said to be the *magnitude* of a body, what does it mean? Is it a *number*, or what is it?

(2) What is the difference between *Gravity* and *Specific Gravity*?

(3) In Tables *of Specific Gravities* what is commonly taken for the *unit*, or *standard?*

(4) Can the same substance have a different *Specific Gravity* under different circumstances? Is *water* such a substance? If so, how can it be used as a *standard?*

(5) If M be expressed in cubic feet, and S in terms of the *Specific Gravity* of Water, in terms of what must W be expressed, in order that the equality $W = MS$ may be true?

(6) What is the *datum* necessary for rendering the formula, $W = MS$, practically useful; so that, for instance, knowing the *Specific Gravity* of gold (19·4) you could apply the formula to find the *weight* of a cubic inch of gold?

(7) In Prop. VIII., where is it assumed, that the floating body is of *uniform density?*

(8) In the definition of *Specific Gravity* of a substance, is it assumed, that the substance is of *uniform density?*

(9) In Prop. VIII., if the *whole* body be just immersed, and float there, what conclusion do you draw as to the *Specific Gravities* of the body and fluid?

(10) Would Prop. VIII. apply to an empty *ship* constructed wholly of iron, and floating in smooth water?

(11) What becomes of the result in Prop. VIII., if the *Specific Gravity* of the body be *greater than* that of the fluid?

(12) In Prop. IX. what is meant by 'weight *lost*'? Is the weight actually *lost?* In what case will a body '*lose*' its *whole* weight in a fluid?

(13) Why does a man learn to swim better in salt water than in fresh?

(14) In Prop. ix., will the result be affected by the greater or less *depth*, to which the solid is immersed, below the surface of the fluid?

(15) Can the *Specific Gravities* of *fluids*, as well as *solids*, be determined by means of the *Hydrostatic Balance?* If so, how?

(16) In certain specimens of milk, how would you be able to detect those, if any, which had been adulterated with water?

CHAPTER IV.

ELASTIC FLUIDS.

79. PROP. XII. *Air has Weight.*

This is proved by the following experiments:—

The weight of a vessel from which the air has been exhausted is found to be less than when it was filled with air.

Or, if into a vessel already filled with common air more air be *forced*, the vessel will then be heavier than it was before.

Also, if a bladder be weighed in a vessel, from which the air has been exhausted, first when the bladder contains no air, and again after air has been forced into it, a greater weight is required to balance the bladder in the latter case.

Whence it is concluded, that "*Air has Weight*".

The same conclusion seems to follow from the simple consideration, that *Air* is a *fluid*, which according to Definition (Art. 58) is a *material body*, and the property of having *weight* is considered as necessarily belonging to *Matter*. See Art. 6.

80. *Air* is a substance which, besides having *weight*, possesses the property of self-expansion, so that the matter of which it consists is continually striving to occupy a greater space; and this effort to expand, measured by the pressure required to counteract it, is called the *Elastic Force* of the air.

ELASTIC FLUIDS.

81. Prop. XIII. *The elastic force of air at a given temperature varies as the density.*

This is proved by experiment.

Let $ABCD$ be a glass tube, of uniform bore, b, having the legs AB and CD *vertical*,—the end A of the tube open, and the other end D closed.

A quantity of mercury is poured in at the open end A, so as to confine a quantity of *air* in the shorter leg CD; the air is then extracted from the longer leg, by means of an instrument for that purpose, and the mercury stands at different heights, E, F, in the two legs. Through E is drawn the horizontal line Ee, meeting the longer leg in e.

Then, the weight of the mercurial column Fe

= pressure *downwards* at e on a surface b, by Prop. IV.,

= pressure *upwards* at e on a surface b, by Prop. I.,

= pressure *upwards* at E on a surface b, by Prop. II.,

= pressure by the air in DE on the surface b of mercury with which it is in contact;

because, the whole being *at rest*, the pressures upwards and downwards on the same horizontal plane must be equal.

Similarly, when more mercury is poured in, if its surfaces stand at G and H in the two legs, drawing Gg horizontally through G, it follows, that

Weight of the mercurial column $Hg=$ pressure by the air in DG on the portion b of the surface of mercury with which it is in contact.

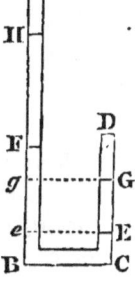

Now, if the lengths ED, Fe, GD, Hg, be measured, it is invariably found, however the quantities of air and of mercury used in the experiment be altered, that (the air retaining the same temperature during the experiment)

$$DE : DG :: Hg : Fe.$$

Now, the Elastic Force of the air in DE is measured, as has been shewn, by the *weight* of the mercurial column

Fe; and that of the air in DG by the *weight* of the mercurial column Hg; therefore, since the *density* of the mercury is uniform,

Elastic Force of the air in DG : that of the air in DE
:: *volume* of mercury in Hg : that of mercury in Fe,
:: *height* Hg : *height* Fe, (since the bore is uniform,)
:: DE : DG, (by what has been shewn,)
:: *content* of tube DE : that of tube DG.

But the *same quantity* of air is in DG which was in DE, and its *density* will be inversely proportional to the space which it occupies*, that is,

Density of the air in DG : that in DE
:: Content of DE : Content of DG,

$\therefore$ Elastic Force of air in DG : that in DE
:: Density of air in DG : that in DE,

or *Elastic Force* of air $\propto$ its *Density*.

82. PROP. XIV. *The elastic force of air is increased by an increase of temperature.*

This is proved by experiment.

If a bladder partially filled with air be brought near the fire, the enclosed air expands as it becomes heated, and the bladder becomes fully distended. As the enclosed air cools down, the bladder becomes more and more flaccid.

---

* This has not yet been proved in this *Course;* but it may be shewn as follows:—Assuming certain standard units of *volume* and *quantity of matter,* the *Density* of a body is most simply defined to be *the quantity of matter in a unit of its volume.* If, then, D be the Density, V the Volume, Q the quantity of matter, of any body or substance,

$$D = \frac{\text{whole quantity of matter}}{\text{whole N}^\circ \text{ of units of volume}} = \frac{Q}{V};$$

and if, while D and V vary, Q remains invariable,

$$D \propto \frac{1}{V}.$$

ELASTIC FLUIDS. 67

83. DEFS. A VALVE is a kind of door which fits an orifice, so that being pressed by a fluid on one side it opens and allows the fluid to pass through, but keeps the orifice tightly closed, if the fluid press on the *other* side.

Valves are of various forms,—a flap of leather (A) fastened at one edge,—a frustum of a cone (B) made of metal,—a sphere (C),—or a plate of metal (D) with an axis passing perpendicularly through it. By any of these contrivances the flow of a fluid *upwards* would be prevented only by the weight of the *valve;* but a rush of fluid from above would carry the *valve* along with it, and keep the orifice which the *valve* fits completely closed.

When the fluids employed are very *rare*,—like air or gas,—the *valves* are generally made of flaps of oiled or varnished silk, which, being attached at two or three points to the surfaces in which the orifices are situated, are raised by very slight pressures, and so allow fluids of exceedingly small densities to pass under them.

84. PROP. XV. *To describe the construction of the common* AIR-PUMP, *and its operation.*

CONSTRUCTION. The AIR-PUMP consists of a glass vessel A, called the *Receiver*, made to fit a table BC so as to be air-tight. A tube DE connects the *Receiver* with a cylinder EF, called the *Barrel*. At the bottom of the *Barrel* there is a valve E opening 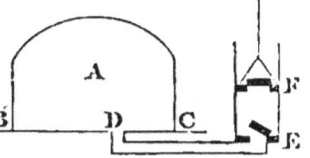 *upwards*, and a piston F (also furnished with a valve opening *upwards*) plays within the *Barrel*.

The instrument is used for pumping the air out of the *Receiver A*.

OPERATION. Suppose the piston F at its *highest* point, and the instrument filled with air the same as that of the surrounding atmosphere.

When the piston is forced down, the air at first in the *Barrel* is condensed, and its elastic force therefore increased (Prop. XIII.)—the valve E is kept closed, and the valve in F being pressed on the under surface more strongly than on

5—2

the upper, opens and allows the air in the *Barrel* to escape through it, until the piston reaches the bottom of the *Barrel*, and the valve *F* closes by its own weight.

Next, on *raising* the piston, the external air keeps *F* closed, and, there now being no air in *EF*, the pressure on the under surface of the valve at *E* will open that valve, and allow air from the *Receiver* and pipe to flow into the *Barrel*, until *F* has reached its highest point. Then the valve *E* closes by its own weight.

(The figure represents the instrument during the *ascent* of the piston.)

When the piston descends again, another barrelful of air escapes through the valve *F*, as before. And so on, until the air in the *Receiver* becomes so rare, that its pressure is insufficient to overcome the weight of the valve at *E*.

Cor. Hence it appears, that although the air in the *Receiver* can be very much rarefied, it cannot be wholly *exhausted*.

85. Prop. XVI. *To describe the construction of the* Condenser, *and its operation*.

Construction. The Condenser is a *Barrel AB*, furnished with a piston *A*, which has a valve in it opening *downwards;* at the bottom of the *Barrel* there is a fixed valve *C*, also opening *downwards*. The neck of the *Barrel* communicates with a strong air-tight vessel *D*, called the *Receiver*.

Operation. Suppose the instrument filled with common air, and the piston at its greatest height. On the piston being forced down, the air in the barrel is condensed, and its elastic force being therefore increased (Prop. xiii.), it keeps the valve *A* closed, opens the valve *C*, and is driven into the *Receiver*. (The figure represents the instrument during the *descent* of the piston.)

On the piston ascending, the elastic force of the air in the *Receiver* closes the valve *C*, and keeps it closed; and there now being no pressure on the under surface of the valve *A*, the pressure of the external air opens that valve, and the *Barrel* becomes filled again with air, which may be driven into the *Receiver* by forcing down the piston, as before. And in this manner the condensation of the air in the *Receiver* may be carried on to any extent required, as far as the strength of the *Barrel* and *Receiver* will permit.

The communication between the *Barrel* and the *Receiver* can be cut off at pleasure by means of a stop-cock at *E*; and the *Barrel* is made to screw off and on at a point above *E*.

86. PROP. XVII. *To explain the construction of the* COMMON BAROMETER, *and to shew, that the mercury is sustained in it by the pressure of the air on the surface of the mercury in the basin.*

The BAROMETER is an instrument for measuring the pressure of the Atmosphere. It consists of a glass tube (see Prop. XVIII.) of uniform bore, closed at one end, and not less than 33, or 34, inches long. This tube is filled with mercury, and the open end, being first stopped with the finger, is placed below the surface of some mercury in a basin, when the tube being fixed in a vertical position and the finger being withdrawn, the mercury in the tube subsides, and stands at a height, above the mercury in the basin, which varies on different days from about 28 to 32 inches. A graduated scale is attached to the upper part of the tube, to mark the height at which the mercury may be standing at any time.

That the column of mercury is supported in the tube *by the pressure of the atmosphere* appears from the experiment, that, when the whole is put into the *Receiver* of an *Air-Pump*, the mercury in the tube *sinks* more and more for every barrel of air that is pumped out, until at length it is all but on a level with the surface of the mercury in the basin[*]. But on readmitting the air into the *Receiver*, the mercury in the tube rises to its original level.

---

[*] It has been shewn, in Prop. XV., that it is not possible to pump *all* the air out of the *Receiver*. If it were so, the surface of the mercury in the tube would then be on *precisely* the same level as the surface in the cistern.

87. Prop. XVIII. *The pressure of the atmosphere is accurately measured by the weight of the column of mercury in the Barometer.*

Let the surface of the mercury in the vertical tube stand at P; and let the area of the section of the tube made by the horizontal surface of the mercury in the basin be represented by AB; CD any portion of area in that surface *equal* to AB; then

Weight of mercury contained in PB

= pressure downwards on the area AB, by Prop. IV.

= pressure upwards on AB, by Prop. I.

= pressure upwards on CD; since AB and CD are equal areas situated in the same horizontal plane of a fluid at rest, by Prop. II.

But the fluid being at rest, the pressure *downwards* on CD, (which arises solely from the pressure of the air in contact with it), is equal to the pressure *upwards* on it.

Wherefore the pressure of the atmosphere on the area CD is equal to the weight of the column of mercury supported in the [vertical] tube of the *Barometer*, of a base AB equal to that area; or "*the pressure of the atmosphere is accurately measured by the weight of the column of Mercury in the Barometer*".

88. Cor. 1. By Art. 64, the pressure of a fluid on a *horizontal* plane immersed in it was shewn to be the weight of a column of the fluid, whose base is equal to the area of the plane, and whose height is the depth of the plane below the surface of the fluid. Wherefore, the pressure exerted by the atmosphere on such an area,—being measured by the weight of the vertical column of mercury (of equal section) which is supported in the *Barometer*,—is equal to the weight of a column of mercury whose base is equal to the plane acted upon, and whose height is the same as that of the column of mercury supported in the vertical tube of the *Barometer*.

89. COR. 2. The pressure of the air being measured by the weight of the column of fluid which it supports, the *Barometer* might be filled with any fluid whatever. But mercury being by far the heaviest *fluid* known, the column of mercury required to produce a given pressure is very much *shorter* than if any other fluid were employed.

For example, if it took 30 inches of *Mercury* to balance the pressure of the air, then since Mercury is 13·6 times as heavy as *Water* (see Art. 69), it would take a column of *Water* 13·6 × 30 inches, or 34 feet high, nearly, to produce the same effect.

90. COR. 3. The mercury in the *Barometer* standing at 30 inches, the pressure of the air on a horizontal square inch of surface is equal to the weight of a column of mercury 30 inches long, and whose base is a square inch; i.e. to the weight of 30×1, or 30 cubic inches of mercury.

To find how much this pressure amounts to.

The weight of a cubic foot of Water is 1000 ounces avoirdupois, very nearly,

∴ the weight of a cubic foot of Mercury = 1000 × 13·6 oz. (Art. 69.)

∴ 1, or 12 × 12 × 12 (cubic foot, cubic inches) : 30 (cubic inches) :: 1000 × 13·6 (oz.) : weight (in ounces) of 30 cubic inches of mercury, which ∴ $= \frac{13600 \times 30}{12 \times 12 \times 12}$ ounces = 236 oz. nearly, or $14\tfrac{3}{4}$ lbs.

And the pressure of the air on a square inch will be greater or less than $14\tfrac{3}{4}$ lbs. according as the mercury in the *Barometer* stands at a greater or less height than 30 inches.

It appears, then, that every square inch with which the air at the Earth's surface is in contact is subjected to a pressure of about 14 lbs.; so that a page of a book 6 inches long by 5 inches wide sustains a pressure of about 6×5×14 lbs., or 420 lbs. The reason why the leaf is not torn by this enormous pressure is, that the pressure of the air on one side of it is counterbalanced by that on the other side.

91. COR. 4. In Chapters I. and II. the pressure of the *Air* on the surface of the fluid contained in an open vessel has been left out of consideration; in other words, the experiments there described

were supposed to be made in the exhausted *Receiver* of an *Air-Pump*.

From Art. 89 it appears, that when *water* is the fluid employed, the pressure of the air on its surface is equivalent (when the mercurial *Barometer* stands at 30 inches), to that which would be produced by a head of water 34 feet deep. In estimating, therefore, the pressure on any surface placed at a given depth below the surface of water, this large additional pressure,—amounting to more than 14 lbs. on a square inch of surface (Art. 90), must be taken into account.

92. PROP. XIX. *To describe the construction of the* COMMON PUMP *and its operation.*

CONSTRUCTION. In the *Common Pump* two hollow cylinders AB and BH, whose axes are in the same vertical line, are connected together, and at their junction is fixed a valve B opening *upwards*. The upper cylinder AB is called the "*Body of the Pump*"; and in it a piston C, containing a valve opening *upwards*, plays by means of a rod attached to the end E of a lever EFG, whose fulcrum is F. A spout D is placed just above the highest point to which this piston ascends. The lower cylinder BH, which is called the "*Suction-Pipe*", reaches below the surface H of a well of water.

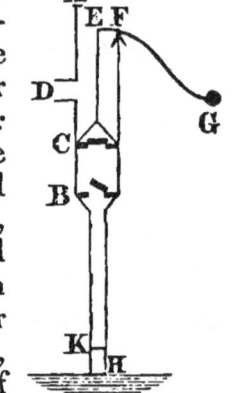

OPERATION. Suppose the piston to be at B, and the *Suction-Pipe* full of common air. As C is raised, the pressure of the external air keeps the valve at C closed, and a vacuum between B and C being consequently made, the air in HB, pressing against the under surface of the valve at B, opens it, and a portion of the air escapes into BC.

The air, therefore, which, before the ascent of the piston occupied the space BH, now occupies the greater space CBH, and so, becoming less dense than before, has less elastic force, and exercises a less pressure on the surface of the water at H, by Prop. XIII. Wherefore, since the *external* air continues to exercise the same pressure as before on the surface of the water in the well, it will force up water into the *Suction-Pipe* to a certain height K, such that the pressure of the air in CK, together with the weight of the column of water KH, produces the same effect on the section of the water in the *Suction-Pipe* at H, as the external air does on an equal area situated in the surface of the water in the well.

When C has reached the highest point of its ascent, and equilibrium exists between the pressure of the external air on the surface of the water in the well on the one hand, and the pressure of the air in CK together with the weight of the fluid column KH on the other, the valve B, being equally pressed on its upper and its under surfaces, will shut by its own weight. The piston C is then pushed down; the air in CB is condensed, until its elastic force becomes greater than that of the external air, when it opens the valve in C and escapes.

When C is raised again, the same circumstances recur. The water rises a little higher in the *Suction-Pipe* at every stroke of the piston, and at last flows through B; and on C descending again, it raises the valve C, passes through it, and on the next ascent of the piston, is brought up to the spout at D.

As the *average* pressure of the atmosphere will not support a vertical column of water more than 34 feet high, if the valve B be more than 34 feet above the surface of the water in the well, the average pressure of the external air not being sufficient in that case to force the water up so high as B, the pump will not work.

N.B. The figure represents the pump during an *ascent* of the piston; when (air, or water, flowing through it) the valve at B is *open*, and that at C is *shut*.

93. PROP. XX. *To describe the construction of the Forcing-Pump, and its operation.*

CONSTRUCTION. The FORCING-PUMP consists of a cylindrical '*Barrel*' AB in which a solid piston C works by means of a rod GC; BD is a '*Suction-Pipe*' reaching below the surface D of a well of water; BE a pipe connecting BC with a vessel EF; at B and E valves are placed, opening *upwards*.

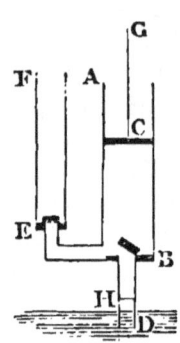

OPERATION. Suppose the piston C to be at its greatest height, the pump full of air, and both valves closed. When C descends, the air in AB is condensed, and its elastic force being increased (Prop. XIII.), the valve E is opened by it, and the greater part of the air that was at first in AB is forced into EF. When the piston reaches its lowest point, the valve E closes by its own weight.

On the piston reascending, the air in EBC being now rendered less dense than that in EF and BD, the valve E is kept closed by the external air, the valve B is opened, and a portion of the air which was at first in DB rushes into the *Barrel*. The external air then forces some water a little way up the pipe (to H suppose), until the pressure on the horizontal section of the *Suction-Pipe* at D (a pressure which arises from the pressure of the rarefied air in HBC together with the weight of the column of water HD), is equal to the pressure of the external air on an equal area in the surface of the water in the well. When this has taken place there is equilibrium; and the valve B, being pressed equally on both its surfaces, closes by its own weight.

The piston again descends, and the air in AB is driven through E into EF; it ascends again, and more water is forced up the *Suction-Pipe:* and these operations are repeated until the *Water* rises above B, and is driven into EF by the next descent of the piston.

As in the case of the common pump, unless BD be less than 34 feet, the pressure of the atmosphere upon the surface of the water in the well will not be sufficient to raise the water above the valve at B; in which case the machine will not work.

N.B. The Figure represents the *Forcing-Pump* during the *ascent* of the piston, when the valve E is *shut*, and the air (or water) is rushing from the *Suction-Pipe* into the *Barrel* through the *open* valve B.

94. Prop. XXI. *To explain the action of the* Siphon.

The Siphon is a bent tube ABC open at both ends, and is often used for drawing fluids out of vessels.

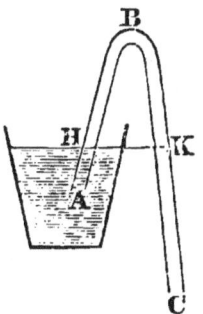

Let the tube be first filled with the fluid, and both ends be then *closed*. Invert the tube, placing one end of it, A, in the vessel of fluid, and so that the other end, C, is *below the surface* of the fluid in the vessel; and let the plane of the surface of the fluid meet the legs of the *Siphon* in H and K, the *vertical* height of B, the highest point of the tube, being restricted to less than the height of a column of the fluid of which the pressure is equal to the pressure of the atmosphere.

If now both ends of the tube be opened, the fluid in it will move in the direction ABC, will flow out at C, and will continue to do so, until the surface of the fluid in the vessel sinks to A. The reason is this:—

The moment A is opened, the pressure within the tube at H will be the atmospheric pressure acting *upwards* upon the column HB; and the column HB, by its weight, which is less than the atmospheric pressure (by supposition), will act in an opposite direction. Similarly, when C is opened, the atmospheric pressure acting upwards at C is opposed by the column BC acting downwards. But the column BH of itself will balance BK; therefore the *Resultant* of all the forces, acting on the system is the effect of the column KC by its weight at C. As there is nothing to counteract this effect, the column of fluid KC will pass out through C; the column BK will follow by its own weight; and the atmospheric pressure on the surface of the fluid in the vessel will prevent a vacuum from being formed at B by forcing up the fluid along HB, so as to keep the column ABC continuous, until the surface of the fluid in the vessel sinks to A.

95. If *Water* be the fluid employed, and the height of B above the surface of the fluid be *greater than* 34 feet, the pressure of the

column *BH* at *H* will be greater than the pressure upwards arising from the pressure of the atmosphere on the surface of the water in the vessel. In that case, therefore, when the end *A* of the siphon is opened, the column *BH* will *sink down* until its extremity stands at a height above *H* in the pipe, such that the weight of the column of water in the tube is just balanced by the pressure of the atmosphere. If *C* be now opened, the water in *CB* will run out at *D*, and (no water from the vessel flowing over *B*), the air will rush up the pipe *CB*, and press on the upper end of the column of fluid supported in *BH*. This column, therefore, being now equally pressed by the atmosphere at each of its extremities, will descend by its own weight into the vessel. THE SIPHON *therefore will not act, when* WATER *is the fluid used, if the vertical height of B above the surface of the fluid in the vessel be not less than* 34 *feet.*

96. PROP. XXII. *To shew how to graduate a* COMMON THERMOMETER.

The THERMOMETER is an instrument for comparing the temperatures (*i. e.* the intensities of *heat*) in solids or fluids.

It consists of a slender glass tube of uniform bore, *closed* at the upper end, and terminating at the lower in a bulb. The bulb and part of the tube are filled with Mercury, Spirits of Wine, or any other fluid (not of a gaseous form) which expands on being heated, and contracts with cold; the remaining part of the tube is a vacuum. If the fluid in the *Thermometer* occupy the same space when the instrument is plunged into two different fluids, the *temperature* of those fluids must be the same.

To graduate a Thermometer. Keeping the tube *vertical*, plunge the bulb, and the part of the tube occupied by the mercury, into *melting snow;* and make a mark *A* at the point to which the mercury falls in the tube. (See Fig. Prop. XXIII.) This is called *the freezing-point.* Next plunge the bulb into *boiling water,* and make another mark *B* at the point to which the mercury rises*. This

---

* The temperatures of *melting snow*, and of *boiling water*, are taken to determine the fixed points *A* and *B* in the scale of the *Thermometer*, because it is found, by experiment, that these temperatures are *fixed* and *invariable* —the mercury *always* falling to *A*, when plunged into melting snow,

ELASTIC FLUIDS. 77

is called *the boiling-point*. The distance AB may then be graduated by equal divisions. In the *Centigrade Thermometer* it is divided into 100 equal parts, called degrees—the *freezing-point* being marked 0°, and the *boiling-point* 100°. In *Fahrenheit's Thermometer* the same length AB is divided into 180 parts—the *freezing-point* being marked 32°, and the *boiling-point* 212°.

97. PROP. XXIII. *Having given the number of degrees on Fahrenheit's Thermometer, to find the corresponding number on the Centigrade Thermometer.*

In *Fahrenheit's* graduation of the scale of the *Thermometer*, (which is that generally made use of in England), 32 is the number placed opposite to the *freezing-point* A; and AB being divided into 180 equal parts (called *degrees*), the number placed opposite to the *boiling-point* B is $32 + 180$, or 212.

In the *Centigrade* thermometer, (which is that generally used on the Continent,) the graduation begins from A, which is marked 0°, and AB is divided into 100 equal parts, called *degrees*.

Now let there be a *Thermometer*, furnished on one side with a scale graduated according to *Fahrenheit*, and on the other with a *Centigrade* scale. Let the mercury rise to any point D, and let F and C be the number of degrees respectively marked opposite to D on the scales of *Fahrenheit* and the *Centigrade*.

Then, since the tube is uniform,

N°. of degrees of *Fahrenheit* in AB, (180)

: N°. .... AD, $(F-32)$

:: N°. .... of the *Centigrade* in AB, (100)

: N°. .... AD, (C);

whether the snow be melting quickly or slowly,—and *always* rising to B, when plunged into boiling water, whether the fire applied to the vessel containing the water be great or small.

78　　　　　　　　HYDROSTATICS.

$$\therefore 180 \times C = 100 \times (F-32);$$

$$\therefore C = \frac{100}{180} \cdot (F-32) = \frac{5}{9} \cdot (F-32),$$

the number of degrees in the *Centigrade Thermometer* corresponding to F degrees *Fahrenheit*.

Ex. If the mercury in *Fahrenheit's* thermometer stand at 77, then the corresponding number of degrees on the *Centigrade* will be

$$\frac{5}{9} \times (77-32) = \frac{5}{9} \times 45 = 25.$$

98. Similarly, if the number of degrees be given in the *Centigrade Thermometer*, the corresponding number on *Fahrenheit's Thermometer* will be found from the same equation. Thus

$$100 \times (F-32) = 180 \times C,$$

$$F - 32 = \frac{180}{100} \cdot C = C + \frac{4}{5} \cdot C,$$

$$\therefore F = 32 + C + \frac{4}{5} C.$$

99. There are other *Thermometers* besides the *Centigrade* and *Fahrenheit's*. In *Reaumur's* Thermometer the *freezing-point* is marked 0, and the *boiling-point* 80. In *De Lisle's* Thermometer the *freezing*-point is marked 150 and the boiling-point 0. Hence, if C, F, R, L denote the same temperature on the respective Thermometers, it is easily shewn, that

$$\frac{1}{5} C = \frac{1}{9}(F-32) = \frac{1}{4} R = 20 - \frac{2}{15} L.$$

100. It may also be noted here, that although the temperature of melting snow is found to be the same under all circumstances, strictly speaking, that is not the case with boiling water. The temperature of boiling water varies with the *atmospheric pressure;* and this circumstance makes it necessary to observe the height of the *Barometer* at the time of *graduating* a *Thermometer*. See *Hydrostatics by Prof. W. H. Miller*, 4th Edition, *Art*. 92.

Questions on Chapter IV.

(1) If "*air has weight*", why do we never take it into account in weighing articles with a common pair of *scales*?

(2) A bladder filled with air, being taken up in a balloon, *burst* at a certain height. How do you account for this?

(3) Is not the *density* of air *diminished* by an increase of temperature? If so, how does Prop. XIV agree with Prop. XIII?

(4) When a tap is inserted in a *full* barrel of ale, why does the ale sometimes refuse to come out? And in that case, what is the remedy?

(5) What is the use of the *Vent-Peg* in a beer-barrel? Explain its action.

(6) What practically limits the degree of exhaustion of air by the *Air-Pump*?

(7) A piece of wood floats in a cup of water under the *Receiver* of an *Air-Pump*; will it sink deeper, or less deep, when the air is exhausted?

(8) What practically limits the extent to which condensation of air may be carried on by means of the *Condenser?*

(9) Does the labour of working the *Air-Pump* with a single *Barrel* increase, or decrease, as the air is gradually exhausted? And why?

(10) Does the labour of working the *Condenser* increase as the operation goes on? And why?

(11) In the common *Barometer*, what is the object of having the surface of the mercury in the basin considerably larger than that in the tube?

(12) What would be the effect of allowing a small quantity of air to be admitted into the vacuum at the top of the tube of a *Barometer?*

(13) Why is *Mercury* used rather than any other fluid in constructing a *Barometer?*

(14) What are the advantages and disadvantages, of a *Water-Barometer?*

(15) A weight is sustained in air by a thread; will the thread be more strained, or less strained, when the *Barometer* rises?

(16) It sometimes happens, that a *Common Pump*, which will not work, is rendered effective by pouring water into it above the piston. What is the explanation of this?

(17) In the *Common Pump*, if the piston be not able to reach the fixed valve at the top of the suction-pipe, under what conditions will the pump not work?

(18) In the *Common Pump*, if the piston were nearly, but not quite, air-tight, how would this affect the action of the pump?

(19) Is the labour of working the *Forcing-Pump* affected by the size of the pipe, as to bore, up which the water is forced?

(20) Why is it necessary for the piston-rod of the *Forcing-Pump* to be made stronger than that of the *Common Pump?*

(21) When the *Siphon* is in use, what would be the effect of making a small hole at its highest point?

(22) How does the *Siphon* help to explain the phenomena of *Intermittent Springs?*

(23) What would be the inconvenience of having the bore of the tube of a *Thermometer* large?

(24) Why is it necessary to note the height of the *Barometer* at the time of determining the *Boiling-Point* in a *Thermometer?*

(25) Can the heat of *steam* be made to exceed $212°$ Fahrenheit? If so, how?

EXAMPLES AND PROBLEMS,

WITH THEIR

SOLUTIONS.

In the Examination for the Ordinary Degree of B.A., (that is, of those who are not Candidates for Honours,) the University requires the attention of its Students to be directed not only to the *Propositions* in the preceding Chapters, but also to "*such Questions, Applications, and Deductions, as arise directly out of the said Propositions*".

The following Collection will give some notion of the sort of *Questions, &c.* which are to be expected, as well as of the manner in which they are required to be answered.

No general *Rules* can be laid down for the solution of problems in *Mechanics* and *Hydrostatics;* but the following suggestions will be found useful :—

1st. Let the Student make sure that he has a clear perception of the meaning of *Ratio* and *Proportion* (see *Wood's Algebra*, Art. 222,) in order that he may avoid the error of comparing things together which are not *of the same kind*. It is a fatal mistake to compare *weight* with *money*—*length* with *volume*—*time* with *superficial area*—and so on.

2nd. In representing *Force, Weight, Length, Volume,* &c., by numbers, or by letters which stand for numbers, let it be always distinctly expressed what is the *unit* of measure. Thus, if the *weight* of a body be represented by W, let it be stated from the beginning, whether it be *tons*, or *pounds*, or *ounces*, &c. of which the number is W. Or if M be the *magnitude* of a body, let it not be left in doubt, whether the *unit*, of which the number in the body is M, be a cubic *yard*, or a cubic *foot*, or a cubic *inch*, or, &c.

L. C. C.

MECHANICS.

1. Forty cubic inches of a substance weigh 25 lbs., and two cubic inches of another substance weigh 1 lb. Compare the *densities* of the two substances.

By Definition (Art. 8),

density of 1st. substance : *density* of 2nd. substance

$$:: \text{weight of 40 in. of 1st.} : \text{weight of 40 in. of 2nd.}$$
$$:: 25 \text{ lbs.} : 20 \text{ lbs.}$$
$$:: 25 : 20$$
$$:: 5 : 4.$$

2. A cubic inch of a substance weighs 280 grains; and a portion of another substance, of twice the density, weighs 400 grains. Find the magnitude of the latter.

Let x be the number of cubic inches required; then

$$\frac{400}{x} = \text{weight of a cubic inch of latter substance, in grains,}$$

∴ density of 1st. substance : density of 2nd. :: $280 : \frac{400}{x}$,

or, by the question, $\quad 1 : 2 :: 280 : \frac{400}{x}$,

$$2 \times 280 = \frac{400}{x},$$

$$\therefore x = \frac{40}{56} = \frac{5}{7} \text{ cubic in.}$$

3. Two weights of 6 lbs. and 9 lbs., respectively, balance each other at the extremities of a straight lever 10 feet long. Find the position of the *fulcrum*.

Let x be the distance, in feet, of the fulcrum from that end at which the weight of 6 lbs. acts; then $10-x$ is its distance from the other end; and, by Art. 21,

$$6 \times x = 9 \times (10-x),$$
$$\text{or } 6x = 90 - 9x,$$
$$15x = 90, \quad \therefore x = \frac{90}{15} = 6.$$

Hence the *fulcrum* is 6 feet from one end, and 4 feet from the other.

4. Two men of equal height carry 4 cwt. by means of a uniform pole, the ends of which rest on their shoulders. The weight is suspended at a distance of *two-sevenths* of the length of the pole from one of the men; how many pounds does the other man support?

Considering this as the case of a *Lever*, with the shoulder of the man who is the nearer to the weight as *fulcrum*, we have, by Art. 23,

$$\text{weight required} \times \frac{7}{7} = 4 \text{ cwt.} \times \frac{2}{7};$$

$$\therefore \text{weight required} = 448 \times \frac{2}{7} \text{ lbs.} = 128 \text{ lbs.}$$

5. A uniform cylinder, 4 feet long, and of which an inch and a half weighs 3 oz., is kept at rest in an horizontal position by a weight of 4 lbs. attached to one end of it. Find the position of the *fulcrum*.

Since $1\frac{1}{2}$ inches weigh 3 oz., 1 inch weighs 2 oz., and the whole cylinder weighs 48×2 oz., or 6 lbs.

Now, by Art. 20, the cylinder will produce the same effect by its weight as if it were collected at its middle point; therefore supposing it so collected, and taking x the distance, in feet, of the *fulcrum* from the end to which the weight of 4 lbs. is attached, we have the case of a straight *Lever* kept at rest by weights 4 lbs. and 6 lbs. acting in the same direction on opposite sides of the *fulcrum* at distances x feet, and $2-x$ feet, respectively.

So then, by Art. 21,

$$4 \times x = 6 \times (2-x),$$
$$4x = 12 - 6x,$$
$$10x = 12, \quad \therefore x = \frac{12}{10} = 1 \cdot 2 \text{ feet.}$$

6. A certain weight, when it is attached first to one end of a straight *Lever* of the first class, and then to the other, is balanced by half a pound, and 18 oz., respectively. Compare the lengths of the *arms* of the *Lever*.

Let the given weight be P oz., and AB the *Lever* with *fulcrum* C, BC being the shorter *arm*; then

$$P \times AC = 18 \times BC, \quad \ldots \ldots (1),$$
and
$$P \times BC = 8 \times AC, \quad \ldots \ldots (2),$$

∴ dividing (1) by (2), $\dfrac{AC}{BC} = \dfrac{18}{8} \times \dfrac{BC}{AC}$,

multiplying by $\dfrac{AC}{BC}$, $\left(\dfrac{AC}{BC}\right)^2 = \dfrac{18}{8} = \dfrac{9}{4}$;

∴ $\dfrac{AC}{BC} = \dfrac{3}{2}$,

or $AC : BC :: 3 : 2$.

7. Two cylindrical rods, one of platina, the other of silver, have the same diameter, and being joined together with their axes in the same straight line, balance in an horizontal position on a fulcrum placed at their point of junction. Given that the length of the platina rod is 9 inches, and that the density of silver is 0·48 times that of platina, find the length of the silver rod.

Let AB, BC, be the lengths of the platina and silver cylinders, respectively; D and E their middle points; and let BC be x inches; then, by Art. 20,

$$\dfrac{BD}{BE}, \text{ or } \dfrac{2BD}{2BE} = \dfrac{\text{wt. of silv. cyl. } BC}{\text{wt. of plat. cyl. } AB},$$

$$= \dfrac{\text{wt. of silv. cyl. } BC}{\text{wt. of plat. cyl. length } BC} \times \dfrac{\text{wt. of plat. cyl. } BC}{\text{wt. of plat. cyl. } AB};$$

but the weights of *equal bulks* of any two substances are proportional to their *densities* (Art. 8); therefore we have

$$\dfrac{9}{x} = \dfrac{48}{100} \times \dfrac{x}{9}, \text{ or } x^2 = \dfrac{81 \times 100}{48};$$

$$\therefore x = \dfrac{9 \times 10}{4\sqrt{3}} = \dfrac{45}{2\sqrt{3}} = \dfrac{15}{2}\sqrt{3}.$$

8. A uniform bent *Lever*, of which the weights of the *arms* are 3 lbs. and 6 lbs. respectively, when suspended on its *fulcrum*, rests with the shorter *arm* horizontal. What weight must be attached to the extremity of the shorter *arm*, that the Lever may rest with its longer *arm* horizontal?

Since the *Lever* is uniform, the lengths of the *arms* will be in the ratio of their *weights*, that is, as 3 to 6, or 1 to 2. So, then, if ACB be the *Lever* with *fulcrum* C, CB is equal to twice AC, AC being the shorter arm.

1st. The *Lever* balances round C with the *arm AC horizontal*. Take D and E the middle points of AC, and CB, respectively; and draw EM vertical to meet AC produced in M. Then, by Art. 27,

$3 \times CD = 6 \times CM$, or $CD = 2CM$, ... (1).

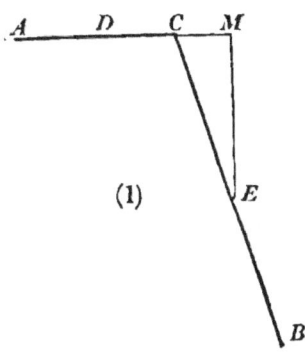

2nd. Let x lbs., acting vertically at A, cause the *Lever* to balance with its *longer arm CB* horizontal; then, constructing as in the annexed fig. (2), it will be easily seen, that the triangle CAP is equal in all respects to the triangle CEM in the former case, (for $\angle ACB$, and ∴ its supplement, remains unaltered; and $CE = CA$); hence $CP = CM$, and

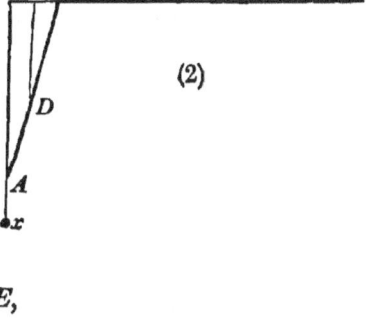

$CN = \frac{1}{2} CP$, ∴ $CD = \frac{1}{2} CA$.

But $x \times CP + 3 \times CN = 6 \times CE$,

or $x \times CM + \frac{3}{2} \times CM = 12 \times CD = 24 \times CM$, by (1),

$$x + \frac{3}{2} = 24, \quad \therefore x = 22\tfrac{1}{2}.$$

9. A uniform heavy rod is supported by two equal strings, attached to its extremities, and meeting together on a fixed peg, so that the rod and strings form an equilateral triangle. Compare the tension of each of the strings with the weight of the rod.

Let AB be the rod, and C the fixed peg; from A draw AE perpendicular to BC; and bisect AB in D. Consider the system as a *Lever AB* with *fulcrum A*, kept in equilibrium by the weight of the rod (W) acting at its middle point D at right angles to AD, and the tension of the string (T) acting in the line BC. Then

$$T : W :: AD : AE, \text{ (Art. 27,)}$$
$$:: \tfrac{1}{2} AB : CD, \ \because AE = CD,$$
$$:: \tfrac{1}{2} AB : \sqrt{AC^2 - AD^2},$$
$$:: \tfrac{1}{2} AB : \sqrt{AB^2 - \tfrac{1}{4} AB^2},$$
$$:: \tfrac{1}{2} AB : \tfrac{1}{2} \sqrt{3 AB^2},$$
$$:: 1 : \sqrt{3}.$$

10. Two forces of $4\tfrac{1}{2}$ lbs. and $3\tfrac{1}{2}$ lbs., applied at a point, have a *Resultant* of 8 lbs. In what directions do the forces act?

Since the *Resultant* is the *sum* of the *component* forces, these latter must act in the same straight line, and in the same direction; in which line, and direction, the *Resultant* also will act.

11. Two forces, which are to each other as 3 : 4, act on a point in directions at right angles to each other, and produce a *Resultant* of 15 lbs. Find the forces.

Let $3x$ and $4x$ represent the number of pounds in the two forces, which produce a *Resultant* of 15 lbs.; then, since the *component* forces act at right angles to each other, they will be represented in magnitude by the sides of a *right-angled* triangle of which the hypothenuse is 15. Hence

$$(3x)^2 + (4x)^2 = (15)^2,$$
$$9x^2 + 16x^2 = 225,$$
$$25x^2 = 225,$$
$$x^2 = 9, \ \therefore \ x = 3;$$

and the required forces are 9 lbs. and 12 lbs.

12. A string, passing round a smooth peg, is pulled at each end by a force of 10 lbs.; and the angle between the parts of the string on opposite sides of the peg is 120°, that is, *two-thirds* of two right angles; find the pressure on the peg both in magnitude and direction.

The pressure required is the *Resultant* of two *equal* forces acting at a point. Let $PA = PB = 10$ lbs. represent the two forces acting at P, $\angle APB = 120°$; complete the parallelogram $PACB$, and

EXAMPLES AND PROBLEMS. 87

draw the diagonal PC; PC is the **Resultant**, or pressure on the peg, both in *magnitude* and *direction*. (Art. 31.)

(1) For *direction*, PC bisects the angle APB, since the forces are *equal*.

(2) For *magnitude*, since $\angle APC = \angle PCB = 60° = \angle BPC$;

therefore remaining angle of the triangle $PBC = 60°$, and the triangle is equilateral. Hence $PC = PA = 10$ lbs.

13. A weight is supported by two strings, which are attached to it, and to two points in a horizontal line. If the lengths of the strings are *unequal*, shew that the *tension* of the shorter string is greater than that of the other.

Let W be the weight; WA, WB, the strings; AB an horizontal line. Draw WC vertical, meeting AB in C; and draw CD parallel to WA, meeting WB in D.

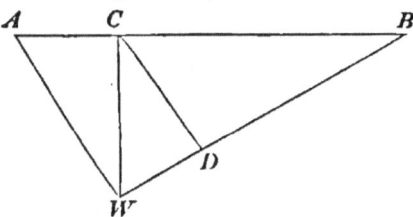

Then W is kept at rest by three forces, its own weight acting in direction CW, tension of string WA, acting in direction DC, and tension of string WB, acting in direction WD. Therefore by converse of Prop. IX., Art. 32, the sides of the triangle WCD represent the *magnitudes* of the forces, respectively. And, therefore,

tension of WA : tension of WB :: CD : WD.

Now, by supposition, $WB > WA$; ∴ $\angle WAB > \angle WBA$; but $\angle BCD = \angle WAB$, ∵ CD is parallel to WA; and BCW is a right angle; ∴ $\angle WCD$, the *complement* of $\angle BCD$, is less than $\angle CWD$, which is the *complement* of $\angle WBA$.

Hence $WD < CD$, (Euclid I, 19,)

i.e. tension of $WA >$ tension of WB.

14. If the *Velocities* of two bodies are as 7 : 2, and the first moves through 42 feet in a second, what space will the other body describe in the same time?

Let the *Velocities* be $7v$ and $2v$; and let x be the space required, *in feet*; then, by Art. 46, Cor.,

$7v : 2v :: 42 : x$,

or $7 : 2 :: 42 : x$;

$$\therefore x = \frac{2 \times 42}{7} = \frac{84}{7} = 12.$$

15. What power will be sufficient to raise a weight of 2 tons, by means of a *Wheel and Axle*, if the radius of the *Axle* : radius of the *Wheel* :: 1 : 10 ?

Let P be the required Power; then, by Art. 37,

$$P : W :: 1 : 10;$$

$$\therefore P = \frac{W}{10} = \frac{2 \text{ tons}}{10} = \frac{20 \text{ cwt.}}{5} = 4 \text{ cwt.}$$

16. In the case of the single moveable pulley, will the *mechanical advantage* be increased or diminished by taking into account the weight of the pulley?

DEF. When a power P balances a weight W on any simple machine, the number of times that W contains P, or the ratio $\frac{W}{P}$, is called the *mechanical advantage* of that machine.

In this case, when the weight of the pulley is neglected, $\frac{W}{P} = 2$, (Art. 41.)

But, if the weight (w) of the pulley be taken into account, it may be supposed concentrated in the centre of the pulley, and acting in the same line as W. Then, reasoning as in Art. 41,

$$P : W + w :: 1 : 2,$$

$$\text{or } W + w = 2P;$$

$$\therefore \frac{W}{P} = 2 - \frac{w}{P}, \text{ which } < 2,$$

and therefore the *mechanical advantage* is diminished.

17. In a system, in which the same string passes round any number of pulleys (Prop. XII), and the parts of it between the pulleys are parallel, if the weight of the pulleys be regarded, under what circumstances will the *mechanical advantage* be reduced to *nothing?*

The weight of the *upper* block will have no effect, because it is supported, by supposition, from without. But the weight (w) of the *lower* block will add to the weight W, and the result, in case of equilibrium, will be

$P : W+w :: 1 : n$, the number of strings at lower block;

$$\therefore W+w=nP,$$

or $\dfrac{W}{P}=n-\dfrac{w}{P}$, which becomes 0, when $w=nP$;

that is, the *mechanical advantage* is nothing, when the weight of the lower block is equal to, or greater than, the *power* multiplied by the number of strings at the lower block.

18. In Prop. XIV, (Art. 45), find the inclination of the plane, when the pressure of W on the plane is equal to P.

In the triangle DEF, DF represents the *power*, FE the *weight*, and DE the *pressure* on the plane; therefore the *pressure* on the plane is equal to the *power*, when $DE=DF$. But $DE : DF :: AC : BC$; therefore the *pressure* $=P$, when

$$AC=BC, \text{ or } \angle ABC = \angle BAC,$$

i. e. when the inclination of the plane equals half a right angle.

19. A weight of 20 lbs. is supported on an *Inclined Plane* by means of a string fastened to a point in the plane; and the string is only just strong enough to carry a weight, hanging freely, of 10 lbs. The inclination of the plane to the horizon being gradually increased, find when the string will break.

As long as there is equilibrium, the tension of the string is represented by P in Prop. XIV; and

$$P : W :: \text{height of plane : its length.}$$

Now, as the inclination of the plane is increased, the height is increased, and P also; and when P exceeds $\frac{1}{2}W$, or 10 lbs., then the string breaks, that is, when the height of the plane first exceeds half its length.

20. Prove that a body may balance in *two* positions round a line, which passes through it, without the *Centre of Gravity* (c. g.) being in that line; but if it balance in *three* positions, the c. g. must be in that line.

If the c. g. be not in the line, the body will nevertheless balance upon the line, when its c. g. is brought into the vertical plane, which passes through the line; and there are *two* such positions—one, when the c.g. is above, and the *other*, when it is below, the line. But if the body be brought into any other position than these two, it will have a tendency to move round the line (Art. 50), unless the c.g. be actually in the line.

21. If, in any triangle ABC, whose c.g. is G, AG be joined, and produced to meet one of the sides in F, shew that $AG = \frac{2}{3}AF$.

Take the fig. in Art. 55, omitting the line bfe, and join EF. Then, since $BE : BA :: 1 : 2 :: BF : BC$, by Euclid VI. 2, EF is *parallel* to AC; and then it is easily shewn, that BFE, and BAC, are similar triangles; and therefore $EF : BF :: AC : BC$, and *alt.* $EF : AC :: BF : BC :: 1 : 2$.

Again, since EF is parallel to AC, it is easily shewn, that EFG, and ACG, are similar triangles, and therefore $AG : AC :: FG : FE$, and *alt.* $AG : FG :: AC : FE :: 2 : 1$; that is, $AG = 2FG$, and therefore $AG = \frac{2}{3}AF$.

22. The *sides*, which contain the right angle, of a right-angled triangle are 3 in. and 4 in., find the distance of the c.g. of the triangle from the vertex of the right angle.

Let ABC be the triangle, $\angle ACB$ a right angle, $AC = 3$ in., $BC = 4$ in.; then $AB^2 = AC^2 + BC^2 = 9 + 16 = 25$; $\therefore AB = 5$ in.

Now bisect AB in D, and join CD; then, with centre D, and radius DA or DB, describe a semi-circle, and it will pass through C, because ACB is a right angle (*Euclid, Bk.* III, 31); therefore

$$DC = DA = \frac{1}{2}AB = 2\tfrac{1}{2} \text{ in.}$$

And, if G be the c.g. of the triangle, it is in CD (Art. 55), and

$$CG = \frac{2}{3}CD, \text{ by last Prob.};$$

$$\therefore CG = \frac{2}{3} \times 2\tfrac{1}{2} = 1\tfrac{2}{3} \text{ in.}$$

23. Weights of 1 lb., 2 lbs., and 4 lbs., being placed at the corners of a given horizontal triangle, which is itself without weight, find the point on which the triangle will balance.

Placing the weights 1 lb., 2 lbs., and 4 lbs. at A, B, and C, respectively, and following the method in Art. 52, the c.g. of the two weights at A and B is at a point D in AB, such that $AD : AB :: 2 : 3$; then joining DC, and supposing the two weights from A and B to act together at D, the c.g. of the *three* weights will be in the line CD, and at a point G such that $DG : CD :: 4 : 3+4 :: 4 : 7$.

24. Shew that the c.g. of three heavy points, of equal weight, and not in the same straight line, coincides with that of the *triangle* formed by joining the points.

Let A, B, C, be the points; then, by Art. 52, since $A = B$, the c.g. of A and B is in the line AB, at a point D equally distant from A and B. Join CD, and suppose A and B collected at D, then the c.g. of A, B, and C, will be in the line CD at a point G such that $CG : DG :: A+B : C :: 2 : 1$, that is, $CG = \frac{2}{3}CD$, and therefore, by Prob. 21, G is the c.g. of the *triangle ABC*.

25. From a square $ABCD$, whose diagonals intersect in O, the triangle AOB is taken away. Find the c.g. of the remainder.

Bisect the sides AD, DC, CB, in E, F, and H, respectively, and join OE, OF, OH. In OE take Oe equal to $\frac{2}{3}OE$; in OF take Of equal $\frac{2}{3}OF$; in OH take $Oh = \frac{2}{3}OH$; and join ef, fh. Then e is the c.g. of the triangle AOD; f is the c.g. of DOC; and h the c.g. of BOC; and the c.g. required will coincide with the c.g. of three equal bodies placed at e, f, and h; that is, with the c.g. of the triangle efh. It will therefore be in the line Of, and at a distance, OG, from O equal to $\frac{1}{3}Of$.

Now $Of = \frac{2}{3}OF = \frac{1}{3}AB$;

∴ dist. of c.g. from $O = \frac{1}{9}AB = \frac{1}{9}$ side of square.

HYDROSTATICS.

[In the following Problems the pressure of the *Atmosphere* is not taken into account, unless the contrary be expressly stated.

Also, it must be constantly borne in mind, that *Water* is the *Unit* of '*Specific Gravities*'; that is, for example, if it be stated, that the s. g. of a particular substance is $10\frac{1}{2}$, the meaning is, that the s. g. of it : the s. g. of water :: $10\frac{1}{2}$: 1, or, in ordinary language, that the particular substance is $10\frac{1}{2}$ times as heavy as *Water*.

Again, it is a fact worth remembering, that, for all practical purposes, the weight of a *Cubic foot* of water may be taken at 1000 oz. *avoirdupois.*]

1. Two rectangular planes, whose lengths are 4 in. and 6 in., and breadths 3 in. and 4 in., are placed horizontally, 4 feet and 2 feet respectively, below the surface of the same fluid. Compare the pressures on them.

The pressures, (by Art. 64), will be in proportion to the superincumbent columns of fluid, that is (since the fluid is supposed to be of uniform density), to the *magnitudes* of those columns. And, the columns being rectangular prisms, their *magnitudes* will be as

the base of the one × its height : base of the other × its height,

$$:: 4\times3\times4 : 6\times4\times2, \text{ that is, } :: 1 : 1.$$

Therefore the pressures are *equal*.

2. If a cubic inch of Mercury weighs 8 oz., what will be the pressure on a square foot of the horizontal bottom of a vessel filled with Mercury to the depth of 6 in. ?

The pressure required = weight of a prism of Mercury, whose base = 144 sq. in., and height = 6 in., that is, weight of 864 *cubic in.* of Mercury. But 1 cubic in. weighs 8 oz.; therefore

the pressure = 864×8 oz. = 432 lbs.

3. Compare the fluid pressures on the top and bottom of a box, 3 feet deep, which is sunk to a depth of 30 feet below the surface of smooth water.

EXAMPLES AND PROBLEMS. 93

The *bottom* of the box being supposed at the depth of 30 feet, and horizontal, and the areas of the top and bottom being the same, the press. on the top : press. on bottom :: depth of top : depth of bottom :: 27 : 30 :: 9 : 10.

But if the *top* be at the depth of 30 feet, then the pressures are
$$:: 30 : 33 :: 10 : 11.$$

4. In two uniform fluids the pressures are the same at depths of 3 in. and 4 in., respectively. Compare the pressures at depths of 7 in. and 8 in., respectively.

Let P_3 represent the pressure in one fluid at depth 3 in.

P_7 .... the same .... 7 in.

p_4 .... the other .... 4 in.

p_8 .... the latter .... 8 in.

Then, by Art. 61, $P_7 : P_3 :: 7 : 3$, $\therefore P_7 = \frac{7}{3} P_3$.

Also $p_8 : p_4 :: 8 : 4 :: 2 : 1$, $\therefore p_8 = 2p_4$.

But $P_3 = p_4$, by the question,

$\therefore P_7 : p_8 :: \frac{7}{3} p_4 : 2p_4 :: \frac{7}{3} : 2 :: 7 : 6.$

5. Shew that the pressure on a square inch in the *side* of a vessel, filled with water, will be between 0·324 lbs. and 0·288 lbs., if the depth of the water be 9 in., and one *side* of the square inch be in the bottom—the weight of a cubic inch of water being 0·036 lbs.

The square inch in question may be in any position between vertical and horizontal. Suppose it vertical, then the highest point in it will be at a depth of 8 in.; and in *any* other position of the square, *every* point in it will be deeper than 8 in. Therefore the pressure on it can in no case be so small as, (reckoning every point in it at 8 in. depth,) 0·036×8 lbs., or 0·288 lbs.

Again, the square inch, being a part of the *side* of the vessel, can never be quite horizontal, for then it would become a part of the *base*. So that the pressure on it can never be so great, as if each point in it were at a depth of 9 in.—that is, can never be as much as 0·036×9 lbs., or 0·324 lbs. The actual pressure, therefore, lies *between* 0·288 lbs. and 0·324 lbs.

6. Find the *Specific Gravity* (s. g.) of earth, when a cubic yard of it weighs a ton.

$$\text{s. g.} = \frac{\text{wt. of a cubic foot of earth}}{\text{wt. of a cubic foot of water}} = \frac{1}{27} \text{ of a ton} \div 1000 \text{ oz.,}$$

$$= \frac{20 \times 112 \times 16}{27 \times 1000} = \frac{35\cdot 84}{27} = 1\cdot 3274 \text{ nearly.}$$

7. The s. g. of a certain fluid is 0·78, and of another 0·66. How much, in weight, of the latter fluid will be of the same bulk as 91 lbs. of the former?

Generally, $W = MS$ (Art. 70); and here, by the question, M is the same for both substances; $\therefore W \propto S$.

Hence $\quad \dfrac{\text{Wt. required}}{91} = \dfrac{0\cdot 66}{0\cdot 78} = \dfrac{66}{78} = \dfrac{11}{13}$;

$$\therefore \text{Wt. required} = \frac{91 \times 11}{13} = 7 \times 11 = 77 \text{ lbs.}$$

8. Twenty pints of a certain fluid weigh exactly the same as 22 pints of another. Compare the s. g.'s of the two fluids.

Here, since $W = MS$, and W is the same in each case, $\therefore MS \propto 1$, or $S \propto \dfrac{1}{M}$.

Hence s. g. of the first : s. g. of the second fluid $:: \dfrac{1}{20} : \dfrac{1}{22} :: 22 : 20 :: 11 : 10$.

9. A quart of water is mixed with a gallon of milk; the s. g. of milk being 1·03, find the s. g. of the mixture.

Let M be the number of cubic inches in a quart of water,
then $4M$ will be .... gallon of milk;
and $5M$ will be .... the *mixture*.

But, since s. g. of water is 1, the weight of the water is $M \times 1$, and the weight of the milk is $4M \times 1\cdot 03$. Therefore the whole weight of the mixture is $M + 4M \times 1\cdot 03$.

Hence s. g. of the mixture = weight of *one* cubic inch,

$$= \frac{M + 4\cdot 12 \times M}{5M} = \frac{5\cdot 12}{5} = 1\cdot 024.$$

10. Five gallons of a fluid, of s. g. 0·98, are mixed with 6 gallons of a fluid of s. g. 1·6, without any *weight* being lost; and the s. g. of the mixture is 1·3. Has *volume* been either gained or lost?

EXAMPLES AND PROBLEMS. 95

Following the same method as in the last Prob., and supposing neither weight nor bulk *lost*,

$$\text{s. g. of the mixture} = \frac{5 \times 0.98 + 6 \times 1.6}{11},$$

$$= \frac{4.9 + 9.6}{11} = \frac{14.5}{11} = 1.318 \ldots$$

which, being *greater* than the actual s. g., shews that volume must have been *gained*.

11. By 20 oz. of a substance being sunk in a cylindrical vessel, whose base is horizontal, and which contains a pint of water, the depth of the water is increased from 4 in. to $4\frac{1}{2}$ in. The weight of a pint of water being 1 lb., find the s. g. of the substance immersed.

The substance displaces a volume of water equal to its own volume, and this volume is equal to a portion of the cylindrical vessel half an inch in depth. Therefore,

volume of water equal in size to the substance : 1 pint :: $\frac{1}{2}$: 4;

and the *weight* of water is proportional to its *volume*,

∴ weight of water equal in size to the substance : 1 lb. :: $\frac{1}{2}$: 4;

$$\therefore \text{weight of water, &c.} = \frac{1 \text{ lb.}}{8} = 2 \text{ oz.}$$

Hence s. g. required $= \frac{20}{2} = 10.$

12. A piece of wood floats with *one-third* of its bulk out of water; and a stone, whose bulk is *one-eighth* that of the wood, when placed upon it, just sinks it below the surface of the water. Find the s. g. of the stone.

Let M be the bulk of the wood in cubic inches; then, by Art. 73,

$$\frac{2}{3}M : M :: \text{s.g. of the wood} : 1, \text{ (s. g. of water)};$$

$$\therefore \text{s. g. of the wood} = \frac{2}{3}.$$

Let s be the s. g. of the stone; then,

$$\text{weight of the stone} = \frac{1}{8}Ms,$$

$$\text{weight of the wood} = M \times \frac{2}{3};$$

and, when the stone is placed on the wood, the weight of the fluid displaced $= M \times 1$. But the two former weights together balance the latter, by Art. 75,

$$\therefore \frac{1}{8} Ms + \frac{2}{3} M = M,$$

$$\text{or } \frac{1}{8} s = 1 - \frac{2}{3} = \frac{1}{3},$$

$$\therefore s = \frac{8}{3} = 2\frac{2}{3}.$$

13. The s. g. of iron is 7·8, and of gold 19·4; find the weight *in water* of a substance composed of 1 lb. of iron and 1 lb. of gold.

The masses of iron and gold being supposed to lose no *bulk* by amalgamation, the whole weight lost in water by the compound will be exactly equal to the sum of the weights lost by the pound of iron and the pound of gold separately considered.

Now the pound of iron loses the weight of an equal bulk of water; but iron is 7·8 times the weight of water, therefore the pound of iron loses $1 \times \frac{1}{7\cdot8}$ lbs. So also the pound of gold loses $1 \times \frac{1}{19\cdot4}$ lbs. Therefore the compound weighs *in water* $2 - \frac{1}{7\cdot8} - \frac{1}{19\cdot4}$ lbs.; that is,

$$2 - \frac{1}{7\frac{4}{5}} - \frac{1}{19\frac{2}{5}} = 2 - 5 \times \frac{97 - 39}{97 \times 39} = 2 - \frac{290}{3783} = 1\frac{3493}{3783} \text{ lbs.}$$

14. An ounce of silver, whose s. g. is $10\frac{1}{2}$, is suspended by a string, so as to be wholly immersed in water. What will be the *tension* of the string?

The *tension* of a string is always measured by the *weight* which it supports, which, in this case, would be 1 oz., if the body were supported *in vacuo*. But the body, by being immersed in water, *loses* the weight of an equal bulk of water, (Art. 75); and therefore the tension of the string is diminished to that extent.

Now, the weight of the silver (1 oz.) $= 10\frac{1}{2}$ times the weight of an equal bulk of water,

$$\therefore \text{weight of an equal bulk of water} = \frac{1}{10\frac{1}{2}} \text{ oz.};$$

hence *tension* required $= 1 - \dfrac{1}{10\frac{1}{2}}$ oz.,

$$= \dfrac{9\frac{1}{2}}{10\frac{1}{2}} = \dfrac{19}{21} \text{ oz.}$$

15. If the water, in the preceding Prob., be contained in a *vessel* placed on a table, will the pressure *on the table* be affected by suspending the silver in it? And how much?

The weight *lost* by the silver is the pressure *upwards* of the water upon it, and is equal to $\dfrac{1}{10\frac{1}{2}}$ oz.

But "fluids press equally in all directions"; therefore the presence of the silver in the water causes an additional pressure *downwards* of $\dfrac{1}{10\frac{1}{2}}$ oz.; and this is communicated to the base of the vessel, and accordingly adds so much to the pressure on the table.

16. If, in the last Prob., we take an ounce of gold, whose s. g. is 19·4, instead of an ounce of silver, will the additional pressure remain the same?

Since the ounce of gold is only about half the *bulk* of the ounce of silver, and the additional pressure in question is the weight of *an equal bulk of water* in each case, it is plain, that the pressure on the table will be much diminished by using gold for silver.

17. A common *Hydrometer* floats in water with 43, and in milk with 100, divisions above the surface. The portion of the stem between each two successive divisions is one 2000th part of the bulk of the whole instrument. Find the s. g. of milk.

Let S be the s.g. of milk, that of water being 1; the *weight* of fluid displaced by the instrument in both cases is the same, being the weight of the instrument; therefore if M be the volume of milk displaced, and M' the volume of water,

$$M \times S = M' \times 1,$$
$$\therefore S : 1 :: M' : M,$$
$$:: 2000 - 43 : 2000 - 100,$$
$$:: 1957 : 1900,$$
$$\therefore S = \dfrac{1957}{1900} = 1\cdot 03, \text{ nearly.}$$

18. Find the weight of a *Hydrometer*, which sinks exactly as deep in a fluid, whose s.g. is 0·9, as it does in water, when loaded with 60 grains.

Let M be the volume of fluid displaced, which, by the question, is the same in both cases; then,

$$W, \text{ (weight of the instrument)} = M \times 0{\cdot}9,$$

and $\quad W + 60 = M \times 1, \because$ s.g. of water is 1;

$$\therefore W + 60 = \frac{W}{0{\cdot}9} = \frac{10\,W}{9},$$

$$9W + 540 = 10W,$$

$$\therefore W = 540 \text{ grains.}$$

19. In the *Common Air-Pump*, if the volume of the *Barrel* be one-fifth of that of the *Receiver*, shew that, after 4 strokes of the piston, the density of the air in the *Receiver* will be about *one-half* of what it was at first.

(By a *stroke* of the piston here is meant a descent and ascent, so that the piston, after going *to the bottom* of the *Barrel*, returns to its original position *at the top*.)

Let B be the content, or volume, of the *Barrel*; then $5B$ is the content of the *Receiver*. And *the same air*, which occupied the space $5B$ at first, will occupy $6B$ after the first stroke. Hence if d be the density of the air at first, and d_1 the density after the first stroke, (by Art. 81, Note,)

$$d_1 : d :: 5B : 6B :: 5 : 6, \quad \text{or } d_1 = \frac{5}{6} d.$$

Similarly, if d_2, d_3, d_4, be the densities of the air in the *Receiver* after 2, 3, 4, strokes of the piston, respectively,

$$d_2 = \frac{5}{6} d_1, \quad d_3 = \frac{5}{6} d_2, \quad d_4 = \frac{5}{6} d_3,$$

$$\therefore d_4 = \frac{5}{6} \times \frac{5}{6} d_2 = \frac{5}{6} \times \frac{5}{6} \times \frac{5}{6} d_1 = \frac{5}{6} \times \frac{5}{6} \times \frac{5}{6} \times \frac{5}{6} d,$$

$$\therefore d_4 = \frac{625}{1296} \cdot d = \tfrac{1}{2} d, \text{ nearly.}$$

EXAMPLES AND PROBLEMS. 99

20. In the last Problem, supposing the pressure of the air at first to be 14 lbs. on the square inch, what is the pressure on a square inch of the inner surface of the *Receiver* after 4 strokes of the piston ?

By Art. 81, the elastic force of air varies as its Density; therefore, in this case, since the density is about one-half its original density, the *pressure* will be about one-half the original pressure, that is, about 7 lbs. on the square inch.

21. Find the pressure of the air in the *Receiver* of a *Condenser* after 10 descents of the piston, the content of the *Receiver* being 10 times that of the *Barrel*.

It is plain, that by 10 descents of the piston, the quantity of air in the receiver is exactly *doubled;* therefore its density is *doubled*, and the elastic force, or pressure, will likewise be *doubled*. So that, if the instrument was filled with common air at first, of which the pressure is about 14 lbs. on the square inch, the pressure of the condensed air will be about 28 lbs. on the square inch.

22. If the atmospheric pressure be 14 lbs. on the square inch, when the *Barometer* stands at 28 inches, what will it be, when the *Barometer* stands at 30 inches ?

By Art. 87, "the pressure of the atmosphere is accurately measured by the weight of the column of mercury in the *Barometer*". And in the same Barometer the weight of the column of Mercury will be proportional to its height. Therefore here,

atmospheric pressure in 1st case : atmospheric pressure in 2nd,

:: 28 : 30,

:: 14 : 15,

or 14 lbs. : required pressure :: 14 : 15,

∴ required pressure = 15 lbs.

23. The height of the mercurial *Barometer* being 30 inches, required the height of a *Barometer*, of which the column above the cistern contains equal weights of mercury (s. g. 13·6), and of proof spirit (s. g. 0·93).

In any *Barometer* the weight of the column in the *vertical* tube is equal to the atmospheric pressure, which in this case is measured by 30 inches of *mercury*. In the compound *Barometer* in question,

7—2

half the column, *in weight*, is mercury; and therefore *its* height is 15 in.

The length of the column of proof spirit is thus found:—

The weight of the column being *equal* to that of 15 inches of mercury, and the thickness of the column the same, since $W = MS$ always, and here $M \propto \dfrac{1}{S}$, the *height* of the column $\propto \dfrac{1}{S}$. Therefore,

height of proof spirit : 15 in. :: 13·6 : 0·93,

$$\therefore \text{height of spirit} = \frac{15 \times 13 \cdot 6}{0 \cdot 93} = \frac{5 \times 13 \cdot 6}{0 \cdot 31} = \frac{6800}{31},$$

$$= 219\tfrac{11}{31};$$

$\therefore$ height of *Barometer* $= 15 + 219\tfrac{11}{31} = 234\tfrac{11}{31}$ inches.

24. Two bodies of different *Specific Gravities* balance in a common pair of scales, when the *Barometer* stands at 28 inches. Will they balance also, when the *Barometer* stands at 30 inches? If not, which will preponderate?

The two bodies, being of different s. G., but equal in *weight* by the first trial, must be unequal in *bulk;* and the greater will be that which has the lesser s. G.

Let M be the number of cubic inches in larger body,

m .... smaller ...

S the s. G. of air, when barometer stands at 30 in.,

s .... 28 ...

In either case the weight lost by the body is the weight of an equal bulk of *air*. The larger body loses MS in one case, and Ms in the other, the difference being therefore $M(S-s)$. The smaller body loses mS in one case, and ms in the other, the difference being $m(S-s)$. And since M is greater than m, the greater body suffers the greater loss of weight by being transferred to air of greater density. Therefore the lesser body will preponderate and the bodies will *not* balance.

25. A *Condenser* at first is full of air, the same as that of the surrounding atmosphere; and the content of the *Barrel* is one-tenth that of the *Receiver*. From a flaw in its construction, the *Receiver* will only sustain a pressure equal to three-fourths that of the atmosphere. During what descent of the piston, and at what part of the descent, will the *Receiver* burst?

Let B, and $10B$, be the contents of *Barrel*, and *Receiver*, respectively;

P the pressure, and D the density, of the atmosphere,

$P_{\prime}$ the pressure, and $D_{\prime}$ the density, of the air in the *Receiver* after x descents of the piston. Then,

$B \times D$ = mass of air forced in by each descent,

and $x \times B \times D =$ .... x descents,

$\therefore 10B \times D + x \times B \times D =$ mass of air in receiver after x descents,

$= 10B \times D_{\prime}$,

$$\therefore D_{\prime} = \frac{10BD + xBD}{10B} = \frac{10+x}{10} . D.$$

But $\dfrac{P_{\prime}}{P} = \dfrac{D_{\prime}}{D}$ (Art. 81), $\therefore P_{\prime} = \dfrac{10+x}{10} . P$; and, by the question, the *Condenser* will burst, when $P_{\prime}$ exceeds $P + \dfrac{3}{4}P$, or $\dfrac{7}{4}P$, that is, as soon as $\dfrac{7}{4}P > \dfrac{10+x}{10} . P$, or $70 > 40 + 4x$, or $x > 7\frac{1}{2}$, that is, in the middle of the *eighth* descent.

26. Shew that 95° of *Fahrenheit's Thermometer* denotes the same temperature as 35° of the *Centigrade*.

By Art. 97 $180 \times C = 100 \times (F - 32)$,

where C and F are the numbers of degrees on the *Centigrade*, and *Fahrenheit, Thermometer*, respectively, for the same temperature.

Here $F = 95$, $\therefore C = \dfrac{100}{180} \times 63 = \dfrac{5}{9} \times 63 = 35$.

27. The *sum* of the numbers of degrees indicating the same temperature on the *Centigrade*, and *Fahrenheit, Thermometer* being equal to 0, find the number of degrees on each.

Here $C + F = 0$, by the question, $\therefore C = -F$.

But $180 \times C = 100 \times (F - 32)$,

$\therefore 180 \times C = 100 \times (-C - 32)$,

$= -100 \times C - 3200$,

$280 \times C = -3200$,

$\therefore C = -\dfrac{320}{28} = -\dfrac{80}{7} = -11\frac{3}{7}$.

And $\therefore F = -C = 11\frac{3}{7}$.

EXAMPLES AND PROBLEMS,

WITH ANSWERS*.

---

i. DENSITY, AND FORCE. (Arts. 8...16.)

(1) Two bodies, each of uniform *density*, are of the same size. Shew that by weighing them it may be determined whether they are of the *same* density.

(2) Eight cubic inches of a substance weigh 5 lbs., and one cubic inch of another substance weighs half a pound. Compare the *densities* of the substances.

(3) A cubic foot of a substance weighs 5 cwt., and a cubic foot of another substance weighs 14 qrs. Compare the *density* of the latter substance with that of the former.

(4) Half a cubic foot of water weighs 500 oz. avoirdupois, and a cubic foot of zinc weighs 7000 oz. Compare the *densities* of water and zinc.

(5) A quart of water weighs 2 lbs., and fifty gallons of proof spirit weigh 372 lbs. Compare the *density* of proof spirit with that of water.

(6) A rectangular log of wood is 16 feet long, $2\frac{1}{2}$ feet wide, $1\frac{1}{4}$ feet deep, and weighs 3000 lbs. Compare the *density* of the wood with that of a metal, a cubic foot of which weighs 480 lbs.

(7) Equal bulks of two substances weigh 4 lbs. and 9 lbs. Find the ratio between two bulks of those substances, which contain the same *quantity of matter*.

(8) Of alcohol 10 pints weigh 8 lbs.; and of oil 10 quarts weigh 18 lbs. Compare the *densities* of alcohol and oil.

* The *Answers* are placed together at the end of the Section.

(9) A body weighs four times as much as another body which is thrice its bulk. Compare their *densities*.

(10) Half a cubic foot of a substance weighs 2 cwt.; what bulk of another substance that is six times as *dense* will weigh 3 cwt.?

(11) A mass of a substance weighs 8 lbs.; what will be the weight of a mass, twice its size, of a substance whose *density* is thrice as great?

(12) A block of wood weighs 60 lbs.; and a piece of iron, one-third its size, weighs 10 st. 10 lbs. Compare the *quantities of matter* in the two substances, and their *densities*.

(13) If a force of 5 lbs. be represented by a line 1 ft. 3 in. in length, what force will a line 2 ft. in length represent?

(14) The measure of a force, when the unit of weight is 1 lb., is 3; what is the measure of the same force, when the unit of weight is 5 lbs.?

---

ii. THE LEVER. (Arts. 17...29.)

[NOTE.—*In the following Examples, unless the contrary be expressly stated, the* LEVER *is supposed to be* straight, *and to rest* horizontally *in equilibrium; and the weights, or other forces, to be applied to it* perpendicularly.]

(1) Half a cwt. acts vertically at the extremity of one of the arms, 10 inches long, of a horizontal straight *Lever*. What number of pounds attached to the extremity of the other arm, 14 inches long, will balance it?

(2) Two weights, of 5 lbs. and 7 lbs. respectively, keep a horizontal straight *Lever* at rest. The length of the arm at which the larger weight acts is 2 feet 1 inch. Find the length of the other arm.

(3) A weight of 14 lbs., suspended 2 inches from the fulcrum of a horizontal straight lever, is balanced by a weight of 8 oz. Find the length of the arm at which the latter weight acts.

(4) A straight *Lever* is kept at rest by two forces acting perpendicularly on it on opposite sides of the fulcrum, and at distances from the fulcrum of 9 inches and 15 inches; the greater force being 6 st. 6 lbs., find the other.

(5) Divide a cwt. into two portions that will balance on a horizontal straight *Lever*, whose arms are 8 feet and 6 feet.

(6) If the length of a *Lever* be 28 in., and the weights, which balance each other at its extremities, are 3 lbs. and 4 lbs., find the position of the *fulcrum*.

(7) On a straight horizontal *Lever*, 20 inches long, two weights, of 24 oz. and of 6 lbs., respectively, balance. Find the position of the fulcrum.

(8) The arms of a horizontal straight *Lever* of the first kind are 12 inches and 18 inches, respectively, and the weight which acts at the extremity of the shorter arm is 3 lbs. Find the pressure on the fulcrum.

(9) On a *Lever* of the first kind 12 lbs. is supported at a distance of 8 inches from the fulcrum. The pressure on the fulcrum being 14 lbs., find the *Power*, and the point of its application.

(10) Two forces, of 3 lbs. and 5 lbs., respectively, balance when they act perpendicularly on a straight *Lever* in opposite directions at the distance of a foot from each other. Find the position of the fulcrum.

(11) At one end of a horizontal straight *Lever* of a given length a force of 3 lbs. acts vertically upwards. At what point must a weight of 12 lbs. be placed to produce equilibrium round a fulcrum situated at the other end?

(12) On a *Lever* of the second kind, the weight of 24 lbs. is suspended at a distance of 8 inches from the fulcrum: what must be the length of the arm at which the power, of 64 oz., acts to balance it?

(13) A force of 48 oz. acts on a *Lever* of the second class at a distance of a foot from the fulcrum, and is balanced by another of 24 lbs., both forces acting perpendicularly to the *Lever*. Find the distance from the fulcrum at which the second force acts; and also the pressure on the fulcrum, and its direction.

(14) Two men, A and B, of equal height, carry a weight suspended from a pole, which rests on their shoulders. Where must the weight be placed, so that A may bear exactly twice as much weight as B does?

THE LEVER.

(15) A *Lever* is held in a horizontal position by two supports that are 5 feet apart, and a weight of 10 lbs. is hung at the distance of $3\frac{1}{2}$ feet from one of the supports. Find the pressure sustained by the other.

(16) On a horizontal rod, 45 inches long, whose extremities are supported, where must a weight be placed so that the pressure on the supports may be as 5 to 4 ?

(17) A *Lever*, 7 feet long, is supported in a horizontal position by props placed at its extremities. Where must a weight of 2 qrs. be placed on it so that the pressure on one of the props may be 8 lbs. ?

(18) The pressure on the *fulcrum* of a horizontal straight *Lever* of the first class is 32 lbs., and the difference of the lengths of the arms is 6 inches; one-eighth, also, of the pressure on the fulcrum is equal to half of one of the weights which balance each other. Find the weights, and the lengths of the arms at which they act.

(19) Two forces, of 8 lbs. and 10 lbs., respectively, acting vertically balance on a horizontal straight *Lever* of the third order; the difference of the lengths of the arms being 2 inches. Find the length of the arm at which the *weight* acts.

(20) A *heavy* uniform rod, 12 feet in length, and weighing 24 lbs., rests horizontally on two props distant 2 feet and 4 feet from the two ends. Find the pressure on each prop.

(21) The arms of a false *Balance* are equal in length, but one *scale* is loaded. Find the true weight of a body by means of such a *Balance*.

(22) If a body be weighed successively at the two ends of a false *Balance*, whose arms are of unequal length, find the true weight of the body.

(23) A weight, when it is attached first to one extremity of a straight *Lever* of the first class, and then to the other, is balanced by half a pound, and 18 oz., respectively. Compare the lengths of the *arms* of the *Lever*.

(24) In a *Lever*, where the Power and the Weight act perpendicularly on the same side of the fulcrum, the Weight is given, and its point of application, and there is no pressure on the fulcrum. Find the *Power*, and the point of its application.

(25) Explain to what class of *Levers* the handles and body of a Wheelbarrow belong.

(26) If a straight *Lever*, from the extremities of which weights are suspended, balance in *any one* position round a fulcrum, it will balance in *every other* position round that fulcrum. Prove this.

(27) A uniform rod, 14 inches long, and 10 lbs. in weight, is joined so as to be in the same straight line with another uniform rod 16 inches long, and weighing 8 lbs. Find the point on which they will balance.

(28) Two forces, acting perpendicularly to a straight *Lever* on the same side of the fulcrum, keep it at rest. The difference of the lengths of the *arms* is 7 inches, and their sum 63 inches. The greater force being $2\frac{1}{2}$ cwt., find the number of pounds in the other.

(29) Two forces, 3 lbs. and 5 lbs., acting perpendicularly to a *Lever* on the same side of the fulcrum, balance. Half the sum of their distances from the fulcrum being 8 inches, find the lengths of the *arms*.

(30) The difference between two weights which keep a horizontal *Lever* at rest is 5 lbs., and the pressure on the fulcrum is 31 lbs. The difference of the lengths of the arms is $4\frac{1}{2}$ inches. Find the weights, and the distance between the points at which they are applied.

(31) At one end of a *Lever*, 20 inches long, a weight of 4 lbs. is placed, and is balanced by a weight at the other end. The sum of the weights, together with the pressure on the fulcrum, being 30 lbs., find the second weight, and the distance of the fulcrum from it.

(32) Two forces, acting perpendicularly and on the same side of the fulcrum, keep a *Lever* at rest. The distance between their points of application is 4 feet 8 inches, the difference of the forces 16 lbs., and the sum of the forces, together with the pressure on the fulcrum, is 60 lbs. Find the forces, and the distance from the fulcrum at which the lesser force acts.

(33) A horizontal uniform cylinder, of which a portion 6 inches long weighs 8 lbs., is $1\frac{1}{2}$ yards in length. A weight of 18 lbs. being attached to one of its extremities, find the position of the *fulcrum* upon which the whole will balance.

THE LEVER.

(34) A heavy uniform rod rests horizontally between two smooth pegs, which are 1 in. apart. The length of the rod is 1 ft., and its weight is 4 lbs. Find the pressure on each peg, one peg being at one end of the rod.

(35) Two cylinders, each of uniform density and of the same diameter, when joined with their axes in the same straight line, balance horizontally on a fulcrum placed at their junction. Their lengths are 9 and 16 inches respectively; and 5 inches of the shorter cylinder weigh 8 lbs. How many ounces will 10 inches of the other cylinder weigh?

(36) There is a cylinder 2 feet long, and whose base is a square inch. From one of its ends a hole is bored $\frac{3}{4}$ of a square inch in area, its axis being the same as that of the cylinder. What must be its length, so that the solid may balance horizontally on a fulcrum placed under the point where the bore ends?

(37) There is a rectangular parallelopiped, 18 inches long, whose ends are squares of which a side is an inch. By taking shavings, each a hundredth part of an inch thick, from two-thirds of the length of one of the faces, the body is made to balance on the line from which the shavings begin. Find the number of the shavings.

(38) Two uniform prisms, whose bases are similar and equal, are joined with their axes coincident; and a fulcrum is placed at their junction. The length of one of the prisms is 8 inches, and 2 inches of it weigh 3 lbs. The length of the other is 13 inches, and 3 inches of it weigh 2 lbs. Shew, 1st, which of the arms will preponderate; and, 2nd, what length must be cut off that arm and joined to the extremity of the other, so that equilibrium may take place round the fulcrum.

(39) The Weight (W) being given in each of the three kinds of *Levers*, shew within what limits the magnitude of the pressure on the fulcrum will lie.

(40) Two cylinders of the same diameter, whose lengths are 1 ft. and 7 ft., respectively, and whose weights are in the ratio of 5 to 3, are joined together, so as to form one cylinder; find the position of the *fulcrum*, on which the whole will balance.

(41) A uniform iron bar is $1\frac{1}{2}$ ft. in length, and 4 lbs. in weight. Find the weight, which acting at one end of the rod, will keep it at

rest in an horizontal position, upon a fulcrum 3 in. distant from that end.

(42) The directions of two forces, P and Q, which act on a bent *Lever*, and keep it at rest, make equal angles with the arms of the *Lever*, which are 6 and 8 inches respectively. Find the ratio of Q to P.

iii. COMPOSITION AND RESOLUTION OF FORCES.
(Arts. 30...34.)

(1) Three forces, of 4 lbs., 2 lbs., and 3 oz. respectively, act upon a point in the same direction, and in the opposite direction forces of 8 oz., and 5 lbs., act. What other single force will keep the point at rest?

(2) Shew that the *Resultant* of two forces, which act on a point in directions not in the same straight line, must be less than the *sum* of the two forces, and greater than their *difference*.

(3) Three *equal* forces act on a point and keep it at rest. Find the angle at which one is inclined to another.

(4) Find the magnitude, and the direction, of the *Resultant* of two given equal forces, which act on a point *at right angles* to each other.

(5) If the sides AB, CA, of a triangle ABC represent in magnitude and direction two forces, which act on the point A, what line will represent the magnitude and direction of their *Resultant?*

(6) If two sides, AB, BC, of a triangle ABC, represent in magnitude and direction two forces which act on the point C, what line will represent their *Resultant* in magnitude and direction?

(7) Can three forces, represented by the numbers 5, 6, and 12, keep a point at rest?

(8) Two forces, of 5 lbs., and 12 lbs., act at right angles upon a point; find the magnitude of their *Resultant*.

(9) Forces represented in magnitude and direction by the sides AB, AC, BC, of the triangle ABC, act upon the point C. Draw

the line which will represent their *Resultant* in magnitude, and in line of action.

(10) Shew that, if at a point A *four* forces be applied, which are represented in magnitude and direction by the sides of the quadrilateral figure $ABCDA$ *taken in order*, the point will remain at rest.

(11) Prove that, if three forces acting on a point keep it at rest, when their intensities are *doubled* they will still keep it at rest, if their directions be not changed.

(12) Given the lines AB, AC, which represent in magnitude, and in line of action, the *Resultant* and one of the two forces (acting at A) of which it is compounded, draw the line which represents the other component force in magnitude, and in line of action.

(13) Forces represented in magnitude and direction by the sides AB, CB, CA, of the triangle ABC, act on the point A. Required to draw the line which will represent the magnitude, and the line of action, of the force that will keep them at rest.

(14) Let ABC be a triangle, and D the middle point of BC. If the three forces represented in magnitude and direction by AB, AC, DA, act upon the point A, find the magnitude and direction of the *Resultant*.

(15) A force of 10 lbs., acting perpendicularly on the arm of a *Lever*, keeps it at rest. This force is removed, and the equilibrium is maintained by another force, applied at the same point in a line which makes half a right angle with the *Lever*: find its magnitude, and shew what other effect it has, besides preventing the *Lever* from moving round the fulcrum.

(16) If a weight be supported by two strings tied to it, which are pulled in directions at right angles to one another by forces of 3 lbs., and 4 lbs., find the weight.

(17) Three forces acting on a point keep it at rest; and they continue to do so, when each force is increased by 1 lb., the directions of the forces remaining as before. What is the inference to be drawn as to the magnitudes of the forces?

(18) A string passing round a small smooth peg is pulled at each end by a force of 10 lbs., and the angle between the two

parts of the string is a right angle. Find the pressure on the peg, and the direction in which it acts.

(19) A string, passing round a smooth peg, is pulled at each end by a force equal to the strain upon the peg. Find the angle between the two parts of the string.

---

iv. VELOCITIES. (Art. 46.)

[*In the following Examples, unless it be intimated otherwise, the motion is supposed to be uniform.*]

(1) A railway train goes at the rate of 30 miles an hour. What is its *Velocity*, estimated by the number of *feet* traversed in a second?

(2) A railway train travels over 150 miles in 5 h. 40 m. What is its average *Velocity* in feet per second?

(3) Compare the *Velocities* of two bodies, of which one moves through 13 yards in $2\frac{1}{4}$ seconds, and the other through 260 feet in a minute.

(4) A body moves with a *Velocity* of 30. With what *Velocity* must another body move, which starts from the same point 3 minutes after the former, and overtakes it in 10 minutes?

(5) For 6 seconds a body moves with a *Velocity* of 10 feet; and for the next 9 seconds with a *Velocity* of 15 feet. What *uniform Velocity* would have carried it over the same space in the same time?

(6) In 12 minutes a man walks a mile. For the first 3 minutes his *Velocity* is 5, and for the last 5 minutes it is 10; what is his *Velocity* during the rest of the time?

---

v. MECHANICAL POWERS. (Arts. 35...48.)

(1) What is the greatest *Weight*, which can be supported by a *Power* of 40 lbs., by means of a '*Wheel-and-Axle*', when the diameter of the Wheel is 10 times that of the Axle?

(2) A *Power* of 56 lbs., by means of a '*Wheel-and-Axle*' supports a weight of 10 cwt.; and the diameter of the *Axle* is 6 inches: what is the diameter of the wheel?

(3) What is the '*Mechanical Advantage*' in the last Prob.? How does it differ from that of another machine, where the diameter of the *Wheel* is 20 in., and that of the *Axle* 1 in.?

(4) In PROP. XI., if the string, by which the *Power* acts, be carried over a *fixed* pulley, so that the *power* may be applied *downwards*, will the '*Mechanical Advantage*' be affected?

Will it be *necessary*, in this case, that the *power* be applied in a direction *parallel to* PA and RB?

(5) In a system of pulleys in which the same string passes round all the pulleys (see PROP. XII.), the '*Mechanical Advantage*' is expressed by the number 10; how many moveable pulleys are there?

(6) In a system in which each pulley hangs by a separate string (PROP. XIII.), the '*Mechanical Advantage*' is equal to 16; how many pulleys are there?

(7) In PROP. XIII., if there be *three* moveable pulleys, and the weight (w) of each be taken into account, shew that

$$P-w : W-w :: 1 : 8.$$

(8) In PROP. XIII., if there be *three* moveable pulleys, find the result of increasing P by 1 oz., and W by 10 oz.

(9) In PROP. XIII., if there be *two* pulleys, equal in weight, shew that P is greater than it would be, if the pulleys were supposed without weight, by a quantity equal to *three-fourths* of the weight of either pulley.

[*In the following Problems the* 'Inclined Plane' *is supposed* perfectly smooth.]

(10) The *Weight* sustained at rest on an *Inclined Plane* by a *Power* acting parallel to the plane is 16 lbs.; and the *Pressure* on the plane is 3 lbs. Find the *Power*.

(11) In PROP. XIV., if the '*Inclined Plane*' be capable of moving horizontally, what force will be required to prevent it?

(12) If a *Weight* of 10 lbs. be placed on an '*Inclined Plane*', whose *base* is 16 feet, and *height* 12 feet, and be attached by a string to an equal *Weight* hanging over the top of the *Plane*, find how much must be added to the weight on the *Plane* that there may be equilibrium.

(13) In Prop. XIV., if P be *three-fifths* of W, what will be the relation between W and the *Pressure* on the *Plane?*

(14) Two '*Inclined Planes*' of the same height are placed together, height to height, their *lengths* being 2 feet and 16 inches, respectively; and two *Weights*, resting upon them, balance each other by means of a string passing over a pulley, the portions of the string being parallel to the planes. The weight supported on the longer plane is 42 lbs., find the other weight.

(15) A '*Wheel-and-Axle*' is applied to sustain a weight upon an '*Inclined Plane*', the string being parallel to the plane. Find the condition of equilibrium.

(16) A weight W is supported on a plane inclined to the horizon at an angle of 30° (one-third of a right angle), by a force P acting parallel to the plane: find the relation between P and W.

(17) A weight of 100 lbs. is suspended from the block of a single moveable pulley, and the string by which the power acts is fastened at the distance of 2 feet from the fulcrum of a straight horizontal *Lever*, 5 feet long, of the second class. Find the force to be applied perpendicularly at the extremity of the *Lever* to preserve equilibrium.

(18) In a system of three pulleys, each of which hangs by a separate string (as in Prop. XIII.), the *Power* (P) is a heavy body weighing 25 lbs., which rests on an '*Inclined Plane*', with the string, passing round a smooth peg, parallel to the plane. The height of the plane being *three-fourths* of its base, find W in case of equilibrium.

(19) In the same system as the last, when P is half a cwt., and the *height* of the '*Inclined Plane*' is half its *length*, find W in case of equilibrium.

(20) A heavy body is supported on an '*Inclined Plane*', whose length is 10 times its height, by means of a string which is parallel to the plane, and is fastened to one end of a horizontal straight *Lever*, so that its direction is at right angles to the *Lever;* and at the other end of the *Lever* is placed 20 cwt. The lengths of the arms of the *Lever* are as 7 : 1; find the *weight* supported on the plane.

vi. CENTRE OF GRAVITY. (Arts. 49...57.)

(1) Three weights of 1 lb., 2 lbs., and 3 lbs., are placed along a straight line a foot apart. Find their C. G.

(2) Three weights of 1 lb., 2 lbs., and 3 lbs., are in given positions, but not in the same straight line. Find their C. G.

(3) Two weights of 6 lbs., and 12 lbs., are suspended at the extremities of a uniform horizontal rod, whose weight is 18 lbs. Find the C. G.

(4) Shew that the C. G. of a parallelogram is the point of intersection of its diagonals.

(5) Find the C. G. of four weights, 1 lb., 2 lbs., 3 lbs., and 4 lbs., placed at the angular points of a given square.

(6) Find the C. G. of a set of weights of 1, 2, 3, 4, and 5, lbs., placed at the angular points of a given regular pentagon.

(7) Weights of 2 lbs. each are placed on two adjacent corners, and weights of 4 lbs. each on the other two corners, of a horizontal square, whose side is a foot in length. Find the distance of the C. G. of the whole from either of the lesser weights.

(8) One side, and the C. G. of a triangle being given, construct the triangle.

(9) On a given horizontal base construct a triangle, such that the vertical line through its C. G. shall pass through one extremity of the base.

(10) Find, by construction, the C. G. of a triangle ABC, and of a heavy body placed at A, one of its angular points, the weight of the heavy body being 7 times that of the triangle.

(11) One half of a given triangle is cut off by a straight line parallel to the base; find the C. G. of the remaining trapezium.

(12) One-third of a parallelogram is cut off by a line parallel to one of its diagonals; find the C. G. of the remainder.

(13) Shew that an isosceles triangle, when placed vertical with one of its equal sides on a horizontal plane will not fall over, whatever be the length of its sides.

(14) Why does a butcher's boy, in delivering meat, generally ride with one stirrup-leather shorter than the other?

(15) If a quadrilateral be such, that one of its diagonals divides it into two *equal* triangles, shew that its c.g. is somewhere in that diagonal.

(16) If a triangle, of uniform density and thickness, be suspended from one of its angular points, shew that it will be divided into two *equal* parts by the vertical plane passing through the point of suspension.

(17) If a triangle, of uniform material, when suspended from one of its corners, has its base horizontal, what is the form of the triangle?

(18) In the leaning tower of Pisa, the top overhangs the base by 12 feet; why does it not fall?

HYDROSTATICS.

vii. PRESSURE OF NON-ELASTIC FLUIDS.

(Arts. 60...67.)

[N.B. *For practical purposes a cubic foot of water may be taken to weigh* 1000 *oz. avoirdupois.*]

(1) What is the pressure on the horizontal bottom of a cistern, filled with water, the area of the bottom being 10 square feet, and the depth of the water one foot?

(2) Find the pressure on the horizontal base of a vessel, whose sides are vertical, and which is filled with fluid, to the depth of 6 inches; a cubic inch of the fluid weighing an ounce and a half, and the area of the base being 64 square inches.

(3) A cubic inch of Mercury weighs 8oz., what will be the pressure on the square inch in a vessel of Mercury at a depth of two feet below the surface?

(4) A vessel, with vertical sides and horizontal base, which contains fluid, is placed in the scale of a *Balance*, and is poised by weights in the other scale. A person, without touching the vessel,

then holds his hands in the fluid. Shew whether this has any effect on the equilibrium of the *Balance.*

(5) If a solid be put into a fluid contained in a vessel, whose sides are vertical and bottom horizontal, and *floats,* explain what additional pressure (if any) there will be on the base.

(6) Explain the advantage of *shallow* cisterns over deep ones, *cæteris paribus.*

---

viii. SPECIFIC GRAVITIES. (Arts. 68...78.)

[*In the following Examples, whenever a mixture is supposed to be made of different substances,—fluid or solid,—unless it be expressly stated otherwise, the Weight and the Volume of the Compound are to be considered respectively equal to the* sum *of the Weights, and the* sum *of the Volumes, of the substances employed.*

*Some substances,—Water and Alcohol for instance,—*lose *bulk when mixed together, and so form a mixture that is of a greater S. G. than if there had been no diminution of bulk. If equal weights of Water (S. G.* 1*) and of Alcohol (S. G.* ·825*) were mixed together without bulk being lost, the S. G. of the mixture (*Proof Spirit*) would be* ·904, *and not* ·93, *the S. G. given in the Table, Art.* 69.

In the following Examples, if the S. G. of a substance be given in figures, the figures are the numbers given in the "Tables of Specific Gravities," explained in Art. 69.]

(1) A cubic inch of iron weighs $4\frac{1}{2}$ oz., and a cubic foot of water 1000 oz., avoirdupois. Find the s. g. of iron.

(2) Fourteen cubic feet of granite weigh a ton. Find the s. g. of granite, a cubic foot of water weighing 1000 oz.

(3) If 9 cubic feet of a certain substance weigh 1000 lbs., find its s. g.

(4) The weight of 540 cubic inches of a substance is 10 lbs., that of 360 cubic inches of another is 40 oz. The s. g. of the second substance being 3, what is that of the other?

(5) Ten cubic inches of a substance weigh 4 lbs.; 4 cubic inches of another weigh 10 lbs. How often does the s. g. of the latter substance contain that of the former?

(6) Find the length of an edge of a cubical block of a substance which weighs 2000 tons, the S. G. of the substance being 1·12 times that of water.

(7) The S.G. of tin being 7·2, find how many cubic inches of copper (S. G. 9) will weigh as much as a cube of tin whose edge is 4 inches. Find also the S. G. of the compound metal, formed by fusing the lumps of tin and copper together.

(8) Ninety pounds of silver are of the same bulk as 75 lbs. of a mixed metal. How many times does the S. G. of silver contain that of the mixed metal?

(9) The S. G.s of two substances are as 45 to 81. The weight of 240 cubic inches of the former is 18 lbs.; what will be the weight of 360 cubic inches of the latter?

(10) Six pints of fluid (s. G. 0·925) are mixed with two quarts of water. Find the S. G. of the mixture.

(11) In 50 cubic yards of rock, whose average S. G. is 1·42, there enter 32 cubic yards of a substance whose S. G. is 1·24. Find the S. G. of the remainder of the rock.

(12) Find the S. G. of a mixture composed of equal bulks of two fluids whose S. G.s are 0·9 and 1·2.

(13) Five pints of a fluid (S. G. 1·04), 2 quarts of another (S. G. 0·97), and 4 gallons of a third (S. G. 1·1), are mixed together; find the S. G. of the mixture.

(14) Portions of two fluids, which are not of the same s. d., form a mixture, of which the S. G. is half the sum of the S. G.s of the fluids composing it. Compare the volumes of the fluids employed.

(15) If W, W', W'', be the weights of two substances and of a compound formed of them, whose S. G.s, as given in "Tables of Specific Gravities", are σ, σ', σ'', respectively, shew that

$$\frac{W''}{\sigma''} = \frac{W}{\sigma} + \frac{W'}{\sigma'}.$$

(16) Find the S. G. of a mixture formed by mixing equal weights of water, and of a fluid whose s. G. is 0·825.

(17) Nine pounds of fluid (S. G. 1·05) are mixed with 7 lbs. of water. Find the ratio of the S. G. of the mixture to that of water.

(18) Eight ounces of salt, mixed with 2 quarts of water, (without increasing its volume) make a brine in which an egg just floats. The weight of a pint of water being 1 lb., find the S. G. of the egg.

SPECIFIC GRAVITIES.

(19) There are mixed together 8, 12, and 20 lbs., of three fluids, respectively; and the s.g.s of the first fluid, the second fluid, and of the mixture, are 1·6, 0·8, and 1·2, respectively. Find the s.g. of the third fluid.

(20) A quart of fluid of s.g. 0·8 is mixed with a gallon of fluid of s.g. 1·05, but from a chymical union taking place the volume of the mixture is one hundredth part less than the sum of the volumes of the portions of the fluids composing it, though no *weight* is lost. Find the s.g. of the mixture to four places of decimals.

(21) The s.g.s of platina, iron, and water being 21·5, 7·8, and 1, respectively, find the numbers to three places of decimals which will represent the s.g.s of platina and water, when the s.g. of iron is taken to be 1.

(22) A vessel, whose sides are vertical and base horizontal, contains three quarts of fluid, the depth of which is 10 inches. When a piece of copper of s.g. 8·9 is immersed, the surface of the fluid rises to $11\frac{1}{2}$ inches above the base. Find the weight of the piece of copper, the weight of a pint of water being 1 lb.

(23) A solid, whose weight is 6 lbs., floats on a fluid, so that the weight of the portion above the surface of the fluid is 2 lbs. Compare the s.g.s of the fluid and solid.

(24) A cube of cork, whose edge is a foot, floating in water, vertically sinks to the depth of 2·88 inches. Find its s.g., that of water being 1.

(25) A solid, of 15 cubic inches in volume, and whose s.g. is 0·6, floats on water. What is the content of the portion of it below the surface?

(26) A block of wood of s.g. 0·64 floats on a fluid with 5-9ths of it below the surface; find the s.g. of the fluid.

(27) A prismatic solid, 10 inches long, floats vertically with 3-5ths of its length immersed. Compare the s.g. of the solid with that of the fluid; and if the solid float vertically after having been shortened 4 inches, find what length of it will then be above the surface of the fluid.

(28) A solid, whose weight is 18 lbs., floats on a fluid, whose density is three times that of the solid. Find the weight of the portion of the solid which remains above the fluid.

(29) A cubical block, whose edge is 3 inches, floats with a side 1 inch above the surface; and if an ounce weight be put on the block, it becomes just even with the surface of the fluid. Find the weight of the solid.

(30) The weight of a cubic foot of oak is 75·6 lbs., and when immersed in water it weighs 12·6 lbs. Find its S.G.

(31) When to one of the faces of a cube of wood, whose edge is 1 foot, a plate of copper of the same area is attached, the whole is found to be of the same S.G. as water. The S.G.s of the wood and of the copper being 0·8 and 9, respectively, find the thickness of the plate.

(32) Given that a cubic foot of water weighs 1000 oz., find how many nails, of 24 to the pound, must be driven up to their heads in a cubic foot of wood of S.G. 0·75, so that the whole may be of the same S.G. as water.

(33) A piece of wood, which weighs 3 lbs., floats in water, and the S.G. of the wood : S.G. of water :: 3 : 4. What weight must be placed on the wood so as just to sink it?

(34) The weight of a piece of wood is 45 lbs.; and when it is put into water 3 lbs. are required to sink it even with the surface. Find the S.G. of the wood.

(35) A body floats with 3 lbs. of it above the surface of the fluid; and when another body, half its size, and of S.G. twice as great, is placed upon it, it sinks even with the surface. Find the weight of the former body.

(36) A weight of 4 lbs., when placed on a piece of wood whose S.G. : S.G. of water :: 3 : 5, just causes it to be immersed. Find the weight of the wood.

(37) A block of wood weighs 10 lbs. in air, and when put into a vessel of water it sinks, pressing the bottom of the vessel with a force of 2 lbs. Find what proportion of the block must be hollowed out,—the orifice of the aperture being closed with wood of the same kind,—so that it may float just even with the surface of the water, when it is held down with a force of 2 lbs. Find also the S.G. of the wood.

(38) A cylindrical vessel of horizontal base contains water. If a cubic foot of cork of S.G. 0·24 be allowed to float on the water, find the additional pressure on the base.

(39) A body, whose volume is 250 cubic inches, weighs 6 lbs. in a fluid, 50 cubic inches of which weigh $2\frac{1}{2}$ lbs. What will the body weigh out of the fluid?

(40) The edges of a rectangular block of marble are 4, 7, and 12, feet; and a piece of the marble that weighs 8 oz. in air, weighs $1\frac{1}{5}$ oz. in water. Find the weight of the block, it being given that a cubic foot of lead (S. G. 11·2) weighs 700 lbs.

(41) 95 ounces of gold, and 99 ounces of lead, balance when weighed in water. The S. G. of gold being 19, find 1st the S. G. of lead, and 2nd. the S. G. of an ounce of gold and an ounce of lead melted together.

(42) If the S. G.s of iron, and of gold, be 8 and 19 times, respectively, that of water, find the weight in water of a substance combined of 1 lb. of iron, and 1 lb. of gold.

(43) A piece of iron weighs 12 lbs. in water; and when a piece of wood, which weighs 5 lbs., is attached to it, the two together weigh 9 lbs. in water. Find the S. G. of the wood.

(44) A piece of iron weighs 10 lbs. in air, and when it is attached to a piece of platina (heavy enough to sink it) which has been immersed in mercury and balanced, the weight that restores the equilibrium is 7 lbs. 6 oz. The S. G. of mercury being 13·6, find that of iron.

(45) A body rests, wholly immersed, between two fluids which do not mix, M cubic inches of it being surrounded by the lighter fluid (S. G. s) and N by the heavier (S. G. s'). Find the ratio $M : M + N$, in terms of the S. G.s of the two fluids, and of that of the solid (s'').

(46) In a mixture, formed of volumes x and y of two fluids, whose S. G.s are s and s' respectively, a body floats with a volume M immersed; but in a mixture formed of the volumes x and y of the fluids whose S. G.s are s' and s respectively, it floats with a volume N immersed. Find the ratio of x to y.

(47) A cylinder of metal, of S.G. 12·4, and 14 inches long, floats wholly immersed, with its axis vertical, in a vessel filled with mercury (S. G. 13·6) and water. Find the length of the portion of the cylinder, which is surrounded by the water.

(48) If W, W', be the weights in air, and w, w', the weights in water, of two substances, the former of which is of a greater density than the latter, shew whether $W:w$ be less, or greater, than $W':w'$.

(49) Three masses, of gold, silver, and a compound of gold and silver, each weigh n ounces in air, and p, q, r, ounces, respectively, when immersed in water. Determine the weight of the gold in the compound.

(50) Three masses, of gold, silver, and a compound of gold and silver, weigh respectively P, Q, and R, ounces in air, and p, q, and r, ounces in water. Find the ounces of gold in the compound.

(51) Taking the quantities P, Q, R, and p, q, r, from the last question, shew what is the order of magnitude of the quantities $p : P$, $q : Q$, $r : R$.

(52) There are masses of two metals which weigh 51, and 39, oz. respectively, and a mass containing certain proportions of the metals weighs 45 oz.; and each of the three bodies loses the same weight, when immersed in the same fluid. Find the number of ounces of the first metal, which enters into the compound.

(53) A vessel which contains fluid is balanced in a pair of scales. A heavy body is then suspended in the fluid as represented in the figure to Art. 74. What effect (if any) is produced on the equilibrium of the scales?

(54) A uniform cube floats with a face, of which the diagonal is 10 inches, horizontal, two weights being placed on the face,—one, of 8 lbs., being at one corner of it; find the position of the second weight, which is 20 lbs.

(55) A cubic foot of water, weighing 1000 oz., is put into a vessel which is of the form of two cylinders, placed with their axes vertical and coincident one upon the other, and communicating with each other. The areas of the *bases* of the lower and the upper cylinders are 54 and 24 square inches, respectively; and the *height* of the lower cylinder is 18 inches. Show that the pressure on the base of the vessel is $1546\frac{7}{9}$ oz.

(56) Two vertical tubes, of equal bore, are connected by a horizontal tube 2 inches long. Supposing 12 inches in length of mercury to be poured into one tube, and 26 of water into the other, find the altitudes of the water and the mercury in the two branches, the s. g. of mercury being supposed to be 13 times that of water.

ix. ELASTIC FLUIDS. (Arts. 79...100.)

(1) Explain why a balloon rises, and why the higher it gets the *slower* it rises. Why does it ever cease to rise?

(2) If the mercury in a *Barometer* stand at 30 inches, what is the greatest vertical length of the suction pipe of a *Common Pump*, which will pump up *Mercury*?

(3) A *Common Pump* will raise water 33 feet; how high will it raise *oil* of the s.g. 0·88?

(4) If 13 inches of water be inserted in the tube of a *Barometer* upon the mercury, what will be the altitude of the upper surface of the water, when the *Common Barometer* stands at 30 inches, the s. g. of mercury being 13? And how much will the top of the water fall, when the mercurial *Barometer* sinks 1 inch?

(5) At what height will the *Water Barometer* stand, when the atmospheric pressure is 15 lbs. to the square inch?

(6) The height of the mercurial *Barometer* being 30 inches, required that of a *Barometer* filled with equal *lengths* of mercury and proof spirit, the s.g. of mercury being 13·6, and that of proof spirit being 0·93.

(7) Under what circumstances might the variation of the *Barometer* prevent the working of the *Common Pump*?

(8) If a cubic inch of *Mercury* weighs 7·8 oz., what will be the atmospheric pressure on a square inch, when the *Barometer* stands at $29\frac{1}{2}$ inches?

(9) A piston fits a hollow cylinder, whose height is 7 inches, and base 1 square foot. Supposing the cylinder originally filled with common air, the pressure of the atmosphere to be 14 lbs. on a square inch, and the piston, by its weight, to sink 1 inch in the cylinder, find the weight of the piston.

(10) The tube of a mercurial *Barometer* is vertical, and of uniform bore. On a syringeful of air being introduced into the upper part of the tube the mercury falls 1 inch; and it falls eight-tenths of an inch more when another syringeful is introduced. The mercury in the cistern being kept at the same level throughout,

find the length of that portion of the tube which was originally a vacuum.

(11) A cylindrical tube, $25\frac{1}{2}$ inches in length, closed at one end, is immersed vertically in water, its closed end being in the surface of the water; find the height to which the water will rise in the tube, assuming that a column of 32 feet of water measures the atmospheric pressure.

(12) Find the pressure against the valve, which opens into the *Receiver* of a *Condenser*, after 15 strokes of the piston, when $A=$ content of the *Receiver*, $B=$ content of *Barrel*, and $P=$ the atmospheric pressure.

(13) If the content of the *Receiver* of an *Air-Pump* be three times that of the *Barrel*, what will be the pressure of the air in the *Receiver* after three strokes of the piston, the atmospheric pressure being 15 lbs. on the square inch?

(14) If a *Siphon* be used for drawing off *Mercury*, what is the greatest height at which the bend may be placed?

(15) Is the extreme depth, from which a *Common Pump* can raise water, always the same? or, is it variable? Upon what does it depend?

THERMOMETERS.

(1) Find the degrees of the *Centigrade Thermometer*, which shew the same temperatures as 77°, 23°, and 32° below zero, of *Fahrenheit*.

(2) Find the degrees of *Fahrenheit*, which shew the same temperatures as 50°, 110°, $-10°$, $-20°·5$, of the *Centigrade*.

(3) Shew that 95° of *Fahrenheit* denotes the same temperature as 35° of the *Centigrade*.

(4) "Temperate" is marked on *Fahrenheit's Thermometer* at 56°, find the corresponding number of degrees on the *Centigrade*.

(5) What number of degrees of *Fahrenheit* corresponds with 4° below zero of the *Centigrade?*

THERMOMETERS.

(6) What number of degrees of the *Centigrade* corresponds with *zero* of *Fahrenheit?*

(7) At what temperature will *Fahrenheit* and the Centigrade shew the same number of degrees?

(8) The temperature at one place is 35° by the *Centigrade*, and at another 64° by *Fahrenheit*. Find the difference of these temperatures by *Fahrenheit*.

(9) Find the temperature at which *Fahrenheit* shews as many degrees above *zero* as the *Centigrade* does below *zero*.

(10) A *Fahrenheit's Thermometer* is not graduated higher than the '*Boiling Point,*' and the length of the scale is one-fourth more than the distance between the '*Freezing*' and '*Boiling*' points. Find the degrees of the *Centigrade* which correspond with the lowest graduation on the instrument described.

(11) The '*Freezing*' and '*Boiling*' points on *De Lisle's* Thermometer are marked 150°, and 0°, respectively; at what temperature will the number of degrees on *De Lisle's*, and on *Fahrenheit's*, be equal?

(12) The *sum* of the numbers, which mark the same heat on the *Centigrade*, and *Fahrenheit*, Thermometer, is 102; find the number on each.

(13) The *sum* of the number of degrees indicating the same temperature on the *Centigrade*, and *Fahrenheit*, Thermometer, being equal to 0, find the number of degrees on each.

(14) How many times will the divisions on *Fahrenheit's* Thermometer coincide with those on the *Centigrade* between *Freezing* point and 0° *Fahrenheit?*

ANSWERS

TO THE PRECEDING

EXAMPLES AND PROBLEMS.

i. Density, and Force.

(1) Being equal in size, they will be equal in density, if they are equal in weight; otherwise, not.

(2) As 5 : 4.
(3) As 7 : 10.
(4) As 1 : 7.
(5) As 93 : 100.
(6) As 1 : 8.
(7) As 9 : 4.
(8) As 8 : 9.

(9) As 12 : 1.
(10) $\frac{1}{8}$ of a cubic foot.
(11) 48 lbs.
(12) (1) Ans. 2 : 5.
 (2) Ans. 2 : 15.
(13) A force of 8 lbs.
(14) $\frac{3}{5}$.

ii. The Lever.

(1) 40 lbs.
(2) 2 ft. 11 in.
(3) 4 ft. 8 in.
(4) 3 st. 12 lbs.
(5) 48 lbs. and 64 lbs.
(6) 16 in. from the 3 lbs.
(7) 4 in. from greater wt.
(8) 5 lbs.
(9) 2 lbs., 4 ft. from fulcrum.
(10) 18 in. from greater force.
(11) At a distance from fulcrum of $\frac{1}{4}$th the length of the Lever.
(12) 4 ft.

(13) (1) Ans. 1$\frac{1}{2}$ in.
 (2) Ans. 21 lbs.
 (3) Ans. In the direction of the greater force.
(14) One-third of the length of the pole from A.
(15) 7 lbs.
(16) 20 in. from either end.
(17) 1 ft. from either end.
(18) Weights 8 lbs. and 24 lbs. Arms 9 in. and 3 in.
(19) 10 in.
(20) 8 lbs. and 16 lbs.

ANSWERS.

(21) Half the sum of the apparent weights, when the body is weighed in both scales successively.
(22) The square root of the product of the apparent wts.
(23) As 2 : 3.
(24) $P = W$, and is applied at the same point.
(25) The 2nd.
(26) See Art. 29.
(27) $\frac{1}{3}$ in. from the junction, on the former rod.
(28) 224.
(29) 10 in., and 6 in.
(30) (1) Ans. 13 lbs., and 18 lbs.
    (2) Ans. 27·9 in.
(31) (1) Ans. 11 lbs.
    (2) Ans. $5\frac{1}{3}$ in.
(32) (1) Ans. 14 lbs. and 30 lbs.
    (2) Ans. 8 ft. 9 in.
(33) 21·6 in. from the wt.
(34) 20 lbs., and 24 lbs.
(35) 81.
(36) 16 in.
(37) 75.
(38) (1) Ans. The longer.
    (2) Ans. $\frac{25}{42}$ in.
(39) In 1st, between W and *any* greater weight.
In 2nd, between 0 and W.
In 3rd, between 0 and ∞, that is, as small, or as great, as we please.
(40) On the longer cylinder, 1 ft. from the junction.
(41) 8 lbs.
(42) 3 : 4.

---

iii. Composition and Resolution of Forces.

(1) 11 oz. acting in the direction of the latter forces.
(3) One-third of 4 right angles.
(4) If P be one of the forces, the *Resultant* $= P\sqrt{2}$; and its direction bisects the angle between the two given forces.
(5) CB.
(6) AC.
(7) No.
(8) 13 lbs.
(9) Produce AC to D, so that $CD = 2AC$, CD is the *Resultant* required.
(12) Join BC, and complete the parallelogram $ABCD$, AD is the line required.
(13) From A draw AD parallel to BC, and equal to $2BC$; AD is the line required.
(14) AD.
(15) (1) Ans. $10\sqrt{2}$ lbs.
    (2) Ans. A pressure of 10 lbs. on the fulcrum in direction of the arm.
(16) 5 lbs.
(17) The forces are equal to one another.
(18) (1) Ans. 17·32 lbs.
    (2) Ans. Bisects the angle.
(19) *Four-thirds* of a right angle.

iv. Velocities.

(1) 44.
(2) 38·8 nearly.
(3) As 4 : 1.
(4) 39.
(5) 13.
(6) $5\frac{3}{4}$.

v. Mechanical Powers.

(1) 400 lbs.
(2) 10 ft.
(3) (1) $\frac{20}{1}$.
    (2) No difference in theory; but in practice the latter machine will be wanting in *strength*.
(4) (1) No.
    (2) No.
(5) 5.
(6) 4.
(8) *W* will be *raised*.
(10) 15·71 lbs.
(11) $P \times \frac{AC}{AB}$, acting horizontally.
(12) $6\frac{2}{3}$ lbs.
(13) As 5 : 4.
(14) 28 lbs.
(15) $P : W :: \frac{rad.\ of\ axle}{rad.\ of\ wheel}$: $\frac{length\ of\ plane}{its\ height}$.
(16) $P : W :: 1 : 2$.
(17) 20 lbs.
(18) 120 lbs.
(19) 2 cwt.
(20) 32,000 lbs.

vi. Centre of Gravity.

(1) Between 2 lbs. and 3 lbs.; 8 in. from the latter.

(2) A, B, C, being the positions of the weights, join AB, take BD one-third of AB, join CD, and bisect it in G; G is the c. g. required.

(3) *Five-twelfths* of the length of the rod from the 12 lbs.

(5) Proceed as in Art. 52.

(6) Proceed as in Art. 52.

(7) 10 inches.

(8) Bisect the given side, AB, in C; join CG, and produce it to D, so that $CG = \frac{2}{3}CD$; ADB is the triangle required.

ANSWERS.

(9) AB the base; draw BC vertical; D any point in BC; join AD, and produce it to E, so that $AD = DE$; ABE is the triangle required.

(10) Bisect BC in D, and AB in E. Join AD, CE, intersecting in F. In AF take G such that $AG : AF :: 1 : 8$; G is the c. g. required.

(11) G, g, being the c. g.s of the whole triangle, and of the triangle cut off, join gG, and produce it to H, so that $GH = Gg$; H is the c. g. required.

(12) Draw the diagonals AC, BD, intersecting in E; the cutting line parallel to AC; g, the c. g. of the triangle cut off, will be in BD. In ED take G such that $EG = \frac{1}{2} Eg$; G will be the c. g. required.

(14) Because he has to lean over on the side opposite to his basket to preserve his balance. Otherwise the vertical through his c. g. would fall beyond his base, and he would fall off.

(17) Isosceles.

HYDROSTATICS.

Pressure of Non-Elastic Fluids.

(1) 625 lbs.
(2) 36 lbs.
(3) 12 lbs.
(4) The vessel will preponderate.
(5) A pressure equal to the weight of the solid.
(6) They are subject to *less* *pressure*, when filled with fluid.

Specific Gravities.

(1) 7·776.
(2) 2·56.
(3) 1·778.
(4) 8.
(5) $6\frac{1}{4}$
(6) 40 feet.
(7) (1) 51·2.
    (2) 8.
(8) 1·2 times.
(9) 48·6 lbs.
(10) 0·955.
(11) 1·74.

(12) 1·05.
(13) 1·08.
(14) The volumes are equal.
(16) 0·904.
(17) 112 : 109.
(18) 1·125.
(19) 1·5.
(20) 1·0101.
(21) S. G. of platina $= 2·756$,
S. G. of water $= 0·128$.
(22) 8·01 lbs.
(23) 3 : 2.
(24) 0·24.
(25) 9 cubic inches.
(26) 1·152.
(27) (1) 3 : 5.
(2) 2·4 inches.
(28) 12 lbs.
(29) 2 oz.
(30) 1·2.
(31) $\frac{3}{10}$ of an inch.
(32) 375.
(33) 1 lb.
(34) 0·9375.
(35) 6 lbs.
(36) 6 lbs.
(37) (1) Two-fifths.
(2) 1·25.
(38) 240 oz.
(39) $18\frac{1}{4}$ lbs.
(40) 25,200 lbs.
(41) (1) 11.
(2) $13\frac{14}{15}$.
(42) $1\frac{25}{30}$ lbs.
(43) 0·625.
(44) $7\frac{15}{80}$.
(45) $\dfrac{s' - s''}{s' - s}$.
(46) $Ms' - Ns : Ns' - Ms$.
(47) $1\frac{1}{3}$ inches.
(48) $W : w$ is the lesser.
(49) $n \cdot \dfrac{r - q}{p - q}$ oz.
(50) $P \cdot \dfrac{Qr - Rq}{Qp - Pq}$.
(51) $p : P, r : R, q : Q$, the first ratio being the greatest.
(52) $25\frac{1}{2}$.
(53) The scale, in which the vessel is placed, will preponderate.
(54) In the diagonal which passes through the former weight, and at a distance of 7 in. from that weight.
(55) The altitudes of the water and mercury are 30 in. and 60 in.

Elastic Fluids.

(1) It rises, because its whole weight, including the gas, is less than that of the air, which it displaces. But the pressure upwards diminishes, as the balloon rises, because the density of the air diminishes; so that its ascent will be slower and slower, until the weight of the balloon exactly balances the weight of air displaced, and then the balloon will cease to rise.

ANSWERS. 129

(2) Just less than 30 in.
(3) $37\frac{1}{2}$ feet.
(4) (1) 42 in.
    (2) 1 in.
(5) 34·56 feet.
(6) 56·16 in.
(7) When the height of the fixed valve above the surface of the well exceeds the height of a *Water-Barometer*.
(8) 14·38125.
(9) 21 cwt., *in vacuo*.
(10) 6·2 in.
(11) $1\frac{1}{2}$ in.
(12) $\dfrac{A+15B}{A} \times P.$
(13) $6\frac{21}{64}$.
(14) Height of Mercurial *Barometer*.
(15) No; it varies according to the height of the *Barometer*.

THERMOMETERS.

(1) 25, −5, −$35\frac{5}{9}$.
(2) 122, 230, 14, −4·9.
(4) $13\frac{1}{3}$.
(5) 24·8.
(6) $17\frac{7}{9}$ below zero.
(7) −40°.
(8) The former is the greater by 49° Fahrenheit.
(9) $11\frac{3}{7}$.
(10) 25 below zero.
(11) $96\frac{4}{11}$.
(12) 25, and 77.
(13) −$11\frac{3}{7}$, and $11\frac{3}{7}$.
(14) Three times.

UNIVERSITY EXAMINATION PAPERS.

The following are the *Papers* of Questions actually given in the Examinations for the ordinary degree of B.A. at Cambridge in the years 1860, 1861, 1862, and 1863.

MECHANICS AND HYDROSTATICS.

SATURDAY, *May* 26, 1860.

FIRST DIVISION.—A.

1. Explain clearly what you understand by the *Weight* of a body.

A cubic inch of lead is suspended by a spiral spring, and the consequent elongation of the spring observed. If the experiment were repeated at the Equator, would the elongation be the same or different? Give your reasons.

2. If two forces, acting perpendicularly at the extremities of the arms of any *Lever*, balance each other, they are inversely as the arms.

A grocer buys tea wholesale at the rate of 4s. a-pound, and in weighing it out to his customers uses a *Balance*, the arms of which are as 16 to 15; at what price must he *profess* to sell it per pound, in order that he may make a profit of 20 per cent.?

3. Enunciate the 'parallelogram of forces'; and assuming it to be true, so far as the *direction* of the *Resultant* is concerned, complete the proof.

Explain the nature and action of the forces, which are called into play, when a boat is towed down a river by men or horses pulling at a single rope.

4. Investigate the conditions of equilibrium on the *Wheel and Axle*.

An endless rope, sufficiently rough not to slip, is stretched round two wheels whose radii are as 1 to 3; find the ratio of the radii of the corresponding *Axles*, in order that equal weights, suspended from strings wound round them in contrary directions, may produce equilibrium.

5. Find the Centre of Gravity of a triangle.

A triangular plate of iron weighing 5 cwt. is carried horizontally by 3 men, one at each angular point; find the weight supported by each.

6. Find the pressure at any point in a mass of fluid at rest, having given that the pressure of the air at the surface of the fluid is P.

A town is supplied with water from a reservoir, the level of the water in which is 136 feet above the lowest point of the main supply-pipe; assuming that the height of the *Water-Barometer* is 34 feet, and that the pressure of air on a square inch is 15 pounds, required the pressure which the main must be constructed to bear.

7. Describe the *Common Hydrometer;* and shew how to compare the Specific Gravities of two fluids by means of it.

An *Hydrometer*, used for determining the Specific Gravity of spirits, is constructed in such a way that the zero point of graduation (at the top of the stem) is on the level of the fluid, when it is placed in proof spirit, and that the interval between two successive graduations corresponds to $\frac{1}{1000}$th part of its bulk. If the reading of the instrument, when placed in spirit sold as *proof*, be 27, determine the amount of water unfairly introduced (Specific Gravity of pure alcohol $=0.8$).

8. Describe an experiment for proving that the elastic force of air at a given temperature varies as the density.

A *Barometer* contains a small quantity of air above the mercury; having given that the length of the tube measured from the constant level of the cistern is 32 inches, and that the mercury in it stands at 29·5 inches, when a standard *Barometer* is at 30, obtain a formula for obtaining the true height of the *Barometer* from the observed height generally.

9. Describe the construction of the *Common Air-Pump*, and its operation.

By what contrivance may the degree of exhaustion of the air in the *Receiver* be exhibited to the eye? Why cannot a perfect vacuum be produced by means of this instrument?

10. Explain the method of filling, and graduating, a *Common Mercurial Thermometer*.

How may the height of a mountain be roughly determined by means of the *Thermometer?*

FIRST DIVISION.—B.

1. Define the terms "*Force*", and "*Weight*".

Explain clearly the method of estimating and comparing statical forces.

2. If two weights, acting perpendicularly on a straight *Lever* on opposite sides of the fulcrum, balance each other, they are inversely as their distance from the fulcrum.

A tobacconist buys tobacco wholesale at the rate of 3s. 4d. a pound, and in weighing it out to his customers uses a *Balance*, the arms of which are as 16 to 15; if he *profess* to sell it at 3s. 9d. a pound, what profit per cent. does he really make?

3. Enunciate the 'parallelogram of forces'; and assuming it to be true so far as the *magnitude* of the *Resultant* is concerned, complete the proof.

It is found that if, on a rapid river, a ferry-boat be turned obliquely to the stream and prevented going down the stream by means of chains stretched across from one bank to the other, it will be carried across by the force of the stream alone. Explain this.

4. Investigate the condition of equilibrium when a weight W is supported on an *Inclined Plane* by a force P acting parallel to the plane.

If the force P be the tension of a fine thread, which passes over a small fixed pulley, and is attached to a weight hanging freely, shew that if P be pulled down through a given space, the height of the *Centre of Gravity* of P and W will remain unaltered.

5. When a body is placed on a horizontal plane, it will stand or fall, according as the vertical line through its *Centre of Gravity* falls within or without the '*base*'.

Why is the word *horizontal* introduced in the enunciation of this proposition? With what modifications is it true of any plane?

6. Shew that the pressure on the horizontal bottom of a vessel, filled with fluid, depends merely on its depth below the surface, and not at all on the quantity of fluid contained in it.

A pipe carries rain water from the top of a house to a large tank, the surplus water in which escapes through a valve in the top which rises freely. A weight of 21 lbs. is placed on it, and it is found that the water rises in the pipe to the height of 20 feet before the valve opens. Required its area (assuming that the height of the *Water-Barometer* is 34 feet, and the atmospheric pressure 15 lbs. on a square inch).

7. Describe the *Hydrostatic Balance;* and shew how it may be applied to compare the *Specific Gravities* of two fluids, by weighing the same solid in each.

A piece of copper of *Specific Gravity* 8·85 weighs 887 grains in water, and 910 grains in alcohol; required the *Specific Gravity* of the alcohol.

8. Describe the construction of the *Condenser*, and the mode of its operation.

If the volume of the cylinder be one-fifth the volume of the *Receiver*, find the pressure at any point of the latter after 20 strokes.

9. Explain the construction of the *Common Mercurial Barometer*.

Having given that the *Specific Gravity* of mercury is 13·57, and that the weight of a cubic inch of water is 252·6 grains, find the pressure of the air on a square inch in lbs., when the mercury in the *Barometer* stands at 30·5 inches.

10. Describe the construction of the *Common Pump*.

If the upward movement of the piston be stopped, when the water has risen to a given height (say 16 feet) in the supply-pipe, but has not yet reached the piston, find the tension of the piston-rod (the area of the piston being 4 square inches, and the atmospheric pressure on a square inch being known).

SECOND DIVISION.—A.

1. If two weights P and Q, acting perpendicularly on a straight *Lever* on opposite sides of the fulcrum, balance each other, determine the position of the fulcrum, and the pressure on it.

The scale-pans of a *Balance* are of unequal weight, and its arms consequently also of unequal length, find the true weight of any substance from its apparent weights, when placed in the two scale-pans respectively.

2. If two forces, acting at any angles on the arms of any *Lever*, balance each other, they are inversely as the perpendiculars drawn from the fulcrum to the directions in which the forces act.

Describe the *Balance* known as the *bent-lever Balance*, (in which the weight (P) is always the same), and shew how to graduate it.

3. If three forces, represented in direction and magnitude by the sides of a triangle taken in order, act on a point, they will produce equilibrium.

Two forces, whose magnitudes are $\sqrt{3} \times P$, and P, respectively act at a point in directions at right angles to each other, find the magnitude and direction of the force which will balance them.

4. In that system of *Pulleys*, in which the same string passes round any number of pulleys, and the parts of it between the pulleys are parallel, there is equilibrium, (neglecting the weights of the pulleys), when $P : W :: 1 :$ the number of strings (n) at the lower block.

If $n=6$, find the greatest weight which a man weighing 10 stone can possibly raise by means of such a system without any further machinery.

5. When a body is suspended from a point, it will rest with its *Centre of Gravity* in the vertical line passing through the point of suspension.

Hence shew, how the *Centre of Gravity* of any plane figure of irregular outline may practically be determined.

6. Describe an experimental proof, that, if the pressure at any point of a fluid be increased, the pressure at all other points will be

equally increased. By what short form of words is this property of fluid pressure sometimes described?

In the common *Hydraulic Press*, are the fluid pressures and tendency to break uniform throughout the cylinders?

7. If a body floats in a fluid, it displaces as much of the fluid as is equal in weight to the weight of the body; and it presses downwards, and is pressed upwards, with a force equal to the weight of the fluid displaced.

Does the above proposition contain all the necessary conditions of equilibrium of a floating body?

A uniform cylinder, when floating vertically in water, sinks to a depth of 4 inches; to what depth will it sink in alcohol of specific gravity 0·79?

8. Describe the *Common Hydrometer*.

If the volume between two successive graduations on the stem of a *Hydrometer* be $\frac{1}{1000}$th part of its whole bulk, and it float in distilled water with 20 divisions, and in sea water with 46, above the surface, required the *Specific Gravity* of the latter.

9. Describe the construction of the *Condenser*, and the mode of its operation.

A cylinder, filled with atmospheric air, and closed by an air-tight piston, is sunk to the depth of 500 fathoms in the sea; required the compression of the air, (assume specific gravity of sea water to be 1·027, specific gravity of mercury 13·57, and height of *Barometer* 30 inches).

10. Explain the action of the *Siphon*.

A hollow tube is introduced into the bottom of a cylindrical vessel through an air-tight collar; and a larger tube, of which the top is closed, suspended over it, so as not quite to touch the bottom, consider the effect of gradually pouring water into the cylinder, until it reaches the level of the top of the inverted tube.

SECOND DIVISION.—B.

1. If two forces P and Q, acting perpendicularly on a straight *Lever* in opposite directions, and on the same side of the fulcrum, balance each other, they are inversely as their distances from the

fulcrum; and the pressure on the fulcrum is equal to the difference of the forces.

2. If the adjacent sides of a parallelogram represent two forces acting at a point in direction and magnitude, the diagonal will represent the *Resultant* force in direction.

Two forces act at a point in directions at right angles to one another, the magnitude of one is P, and the magnitude of the *Resultant* is $2P$, find the *direction* of the *Resultant*, and the magnitude of the other force.

3. In that system of *Pulleys*, in which each pulley hangs by a separate string, and all the strings are parallel, there is equilibrium (neglecting the weights of the pulleys) when $P : W :: 1 :$ that power of 2, whose index is the number of moveable pulleys.

If the pulleys were heavy, and the weight of each $= W_1$, find the additional force P_1, which would be necessary to counteract the effect of their weights.

4. If P and W balance each other on the *Wheel-and-Axle*, and the whole be put in motion, $P : W :: W$'s velocity $: P$'s velocity.

Under what conditions is a similar proposition true of other machines? Shew that in all cases when it obtains, no increase in the *work done* is effected by the intervention of the machine, defining carefully the expression in italics.

5. If a body balance itself on a line in all positions, the *Centre of Gravity* is in that line.

Assuming the position of the *Centre of Gravity* of a triangle to be known, shew that the position of the *Centre of Gravity* of any trapezium may be practically determined by means of this proposition.

6. Explain the so-called *Hydrostatic Paradox*.

Describe the construction and method of application, of the *Hydraulic Press*.

If the area of the small cylinder be $1\frac{1}{8}$ inches, and the diameter of the large piston 20 inches, find the lifting power of the machine under a pressure of 1 ton exerted on the piston of the small tube.

7. When a body is immersed in fluid, the weight lost is to the whole weight as the *Specific Gravity* of the fluid is to the *Specific Gravity* of the body.

Hence deduce the conditions of equilibrium of a floating body.

A piece of cork (*Specific Gravity* ·24) containing 2 cubic feet is kept below water by means of a string fastened to the bottom, find the tension of the string.

8. Describe the *Hydrostatic Balance;* and shew how it may be employed to determine the *Specific Gravity* of a body heavier than water.

Two bodies, whose weights are w_1 and w_2 in air, weigh each w in water, compare their *Specific Gravities.*

9. Shew that, in the *Common Barometer*, the weight of that portion of the mercurial column, which is above the free surface of the mercury, accurately measures the pressure of the atmosphere.

A *Barometer* is sunk to the depth of 20 feet in a lake, find the consequent rise in the mercurial column (*Specific Gravity* of mercury $= 13·57$).

10. Explain the action of the *Siphon.*

A *Siphon* is placed with one end in a vessel full of water, and the other in a similar empty one, both of which are on the plate of an air-pump. As soon as the water has covered the lower end of the *Siphon*, a *Receiver* is put on, and the air rapidly exhausted, and then gradually readmitted; describe the effects produced.

SATURDAY, *May* 25, 1861.

FIRST DIVISION.—A.

1. Shew that forces may be properly represented by straight lines.

2. If two weights, acting perpendicularly on a straight *Lever*, on opposite sides of the fulcrum, balance each other, they are inversely as their distances from the fulcrum.

If the pressure on the fulcrum be equivalent to a weight of 15 lbs., and the difference of the magnitudes of the forces to a weight of 3 lbs., find the forces, and the ratio of the arms at which they act.

3. If the adjacent sides of a parallelogram represent two forces acting at a point in direction and magnitude, the diagonal will represent the *Resultant* force in direction.

At what angle must two forces, P, and $2P$, act upon a point, that the direction of their *Resultant* may be at right angles to the direction of one of the forces?

4. Find the condition of equilibrium when a weight W is supported on an *Inclined Plane* by a force P acting parallel to the plane.

If the force acts horizontally, there is equilibrium when P is to W as the height of the plane is to its base.

5. Find the *Centre of Gravity* of three heavy points; and shew that the pressure on the *Centre of Gravity* is equal to the sum of the weights in all positions.

6. When a body is placed on a horizontal plane, it will stand or fall, according as the vertical line drawn from its *Centre of Gravity* falls within or without the *base*.

A board, in the shape of a right-angled triangle, is placed in a vertical plane, with its right angle resting on a rough horizontal floor, and one of its acute angles leaning against a vertical wall perpendicular to the plane of the board; find its position when the pressure on the wall is the least possible.

7. If a body float in a fluid, it displaces as much of the fluid as is equal in weight to the weight of the body.

Why can a man swim on his back more easily than in any other position?

8. Define *Specific Gravity*. What is meant, when the *Specific Gravity* of a substance is said to be 0.00125?

The *Specific Gravity* of sea water being 1.027, what proportion of fresh water must be added to a quantity of sea water, that the *Specific Gravity* of the compound may be 1.009?

9. When a body of uniform density floats on a fluid, the part immersed is to the whole body as the *Specific Gravity* of the body is to the *Specific Gravity* of the fluid.

A solid sphere floats in a fluid with three-fourths of its bulk above the surface: when another sphere half as large again is attached to the first by a string, the two spheres float at rest below the surface of the fluid; shew that the *Specific Gravity* of one sphere is 6 times greater than that of the other.

10. Describe the construction of the *Forcing Pump* and its operation.

What will be the effect of making a small aperture in the *Barrel?* If the piston works uniformly up and down the length of the *Barrel*, and a small aperture be made one-third of the way up the *Barrel*, how much more time than before will be consumed in filling a tank?

11. Describe the method of filling and graduating a *Common Thermometer*.

Shew how to graduate a *Thermometer*, on whose scale $20°$ shall denote the *Freezing-Point*, and whose 80th degree shall indicate the same temperature as $80°$ Fahrenheit.

FIRST DIVISION.—B.

1. Define *Gravity*, and *Weight*. How is Statical Force measured?

2. If two forces, acting perpendicularly on a straight *Lever* in opposite directions and on the same side of the fulcrum, balance each other, they are inversely as their distances from the fulcrum.

If the pressure on the fulcrum be equivalent to a weight of 3 lbs., and the sum of the magnitudes of the forces to a weight of 15 lbs., find the forces, and the ratio of the arms at which they act.

3. If three forces, represented in magnitude and direction by the three sides of a triangle, when taken in order, act upon a point, they will keep it at rest.

Two strings are respectively fastened by their upper extremities to two points A and B in the same horizontal line; the string at A carries a weight W at its lower extremity which is passed through a ring attached to the lower extremity of the string at B; if the distance between the points A and B is so adjusted, that the strings rest at equal inclinations to the horizon, the tension of the string at B is equal to the weight W.

4. Find the condition of equilibrium on the *Wheel-and-Axle*. Why is the labour of drawing a bucket of water out of a common

well generally greater during the last part of the process than during the first?

5. Find the *Centre of Gravity* of two heavy points, and shew that the pressure at the *Centre of Gravity* is equal to the sum of the weights in all positions.

6. When a body is suspended from a point, it will rest with its *Centre of Gravity* in the vertical line passing through the point of suspension.

Two weights, W and $2W$, are connected by a rigid weightless rod, and also by a loose string, which is slung over a smooth peg: compare the lengths of the string on each side of the peg, when the weights have assumed their position of equilibrium.

7. The surface of every fluid at rest is horizontal.

Shew how the inclination of a table to the horizon may be estimated by means of a tube bent into the form of the arc of a circle, and very nearly filled with fluid.

8. Explain fully the meaning of the equation, 'the weight of a body = its magnitude × its *Specific Gravity*.'

If 2 cubic feet of a substance weigh 100 lbs., what is its *Specific Gravity?*

9. When a body is immersed in a fluid, the weight lost is to the whole weight as the *Specific Gravity* of the fluid is to the *Specific Gravity* of the body.

The cavity in a conical rifle bullet is usually filled with a plug of some light wood. If the bullet be held in the hand beneath the surface of water, and the plug then removed, will the apparent weight of the bullet be increased or diminished?

10. Describe the construction of the *Condenser*, and its operation.

If the direction in which the valves open were reversed, into what instrument would the condenser be converted? Describe the effect of making a small aperture in the *Barrel*.

11. Explain the construction of the *Common Barometer*.

Will the actual rise or fall of the mercury in the tube, observed by means of fixed graduations, accurately measure the increase or decrease of the pressure of the atmosphere?

SECOND DIVISION.—A.

1. Define force, and explain how forces are measured. Shew that they may properly be represented by straight lines.

Construct a triangle, whose sides will represent the forces 2·142857 lbs., 2·009 lbs., and 2·009 lbs.

2. If two weights, acting perpendicularly on a straight *Lever* on opposite sides of the fulcrum, balance each other, they are inversely as their distances from the fulcrum; and the pressure on the fulcrum is equal to their sum.

A heavy pole, weighing 12 lbs., whose *Centre of Gravity* falls at a distance of one-third of the pole from one end, is carried horizontally by two men, one at each end, find the weight supported by each man.

3. If three forces, represented in magnitude and direction by the sides of a triangle, act on a point, they will keep it at rest.

$ABCD$ is a quadrilateral inscribed in a circle; forces P, Q, R, acting in directions AB, AD, CA would keep a particle at rest, shew that $P : Q :: CD : BC$.

4. In a system in which each pulley hangs by a separate string, and the strings are parallel, there is an equilibrium, when $p : w :: 1$: that power of 2 whose index is the number of moveable pulleys.

How may a boy, who can only lift 16 stone, be enabled to raise 1024 stone?

5. Define *Velocity*.

If p and w balance each other in the system of *Pulleys*, in which the same string passes round all the pulleys, and the whole be put in motion, shew that $p : w :: w$'s velocity in direction of gravity : p's velocity.

6. When a body is placed upon a horizontal plane, it will stand or fall, according as the vertical line, drawn from its *Centre of Gravity*, falls within or without its base. With what restriction is this true of any plane?

A sugar-loaf stands on an *Inclined Plane*, rough enough to prevent sliding, whose inclination to the horizon is 45°, shew that it will fall over, if the height of the sugar-loaf be more than twice as great as the diameter of its base—the *Centre of Gravity* of a cone being distant from the base one-fourth of the height.

7. How is the pressure at a point in a fluid estimated? Explain the *Hydrostatic Paradox*.

If the area of the larger surface be diminished, the other circumstances of the explanation remaining the same, what would be the effect?

8. Define *Specific Gravity*, and explain how it is measured.

Explain how to find the weight of a body whose *Specific Gravity* is known. Find the weight of 36 cubic inches of cork, whose *Specific Gravity* is 0·24.

9. When a body of uniform density floats on a fluid, the part immersed : the whole body :: the *Specific Gravity* of the body : the *Specific Gravity* of the fluid.

A body, whose weight is 6 lbs., weighs 3 lbs. and 4 lbs. respectively in two different fluids, compare the *Specific Gravities* of the fluids.

10. Describe the construction of the *Common Pump*, and its operation. What limits the height to which water can be raised by a *Common Pump?* How is this obviated by a *Forcing* or *Lifting Pump?*

11. Shew how to graduate a *Common Thermometer*.

One *Thermometer* marks two temperatures by $9°$, and $10°$; another *Thermometer* by $12°$, and $14°$; what will the latter mark, when the former marks $15°$?

SECOND DIVISION.—B.

1. Define *Force;* how are forces measured? Shew that they may properly be represented by straight lines.

Construct a triangle, whose sides will represent the forces $3·076923$ lbs., $3·416$ lbs., and $3·127$ lbs. respectively.

2. If two forces, acting perpendicularly on a straight *Lever* in opposite directions and on the same side of the fulcrum, balance each other, they are inversely as their distances from the fulcrum; and the pressure on the fulcrum is equal to the difference of the forces.

A heavy pole, weighing 12 lbs., whose *Centre of Gravity* falls at a distance of one-fourth of the pole from one end, is fastened at this end by a rope to the ceiling, and the other end is raised by a man so that the pole is horizontal, what weight will the man support?

3. If three forces, represented in magnitude and direction by the sides of a triangle, act on a point, they will keep it at rest.

ABCD is a quadrilateral inscribed in a circle, forces P, Q, R, act in directions AB, AD, CA, so that $P : Q : R :: CD : BC : BD$, shew that P, Q, R, form a system in equilibrium.

4. In a system of *Pulleys*, in which the same string passes round any number of pulleys, and the parts of it between the pulleys are parallel, there is an equilibrium when $p : w :: 1 :$ the number of strings at the lower block.

How may a boy, who can only lift 15 stone, be enabled to raise 90 stone?

5. Define *Velocity*.

If p and w balance in the system of *Pulleys*, in which each pulley hangs by a separate string, and the whole be put in motion, shew that $p : w :: w$'s velocity in direction of gravity $: p$'s velocity.

6. When a body is placed on a horizontal plane, it will stand or fall, according as the vertical line, drawn from its *Centre of Gravity*, falls within or without its *base*. With what restriction is this true of any plane?

A sugar-loaf, whose height is twice as great as the diameter of its base, stands on a table, rough enough to prevent sliding, one end of which is gently raised until the sugar-loaf is on the verge of falling over, when this is the case, find the inclination of the table to the horizon—the *Centre of Gravity* of a cone being distant from the base one-fourth of the height.

7. How is pressure at a point in a fluid estimated? Shew that the surface of every fluid at rest is horizontal.

In supplying a town with water, why is the locality of the reservoir selected in the highest position possible?

8. If a body floats on a fluid, it displaces as much of the fluid as is equal in weight to the weight of the body; and it presses downwards, and is pressed upwards, with a force equal to the weight of the fluid displaced.

A symmetrical box, weighing 8 lbs., with a weight on the top, floats just immersed in a fluid: how heavy must the weight be, in order that, when removed, the box may float with only one-third immersed?

9. Define *Specific Gravity*, and explain how it is measured. Shew how to find the weight of a body whose *Specific Gravity* is known.

Find the weight of 54 cubic inches of copper, whose *Specific Gravity* is 8·8.

10. Describe the construction of the *Common Air-Pump*, and its operation. What advantage does *Hawksbee's* possess over the *Common Air-Pump?*

11. Shew how to graduate a *Common Thermometer*.

One *Thermometer* marks two temperatures by $8°$, and $10°$; another *Thermometer* by $11°$, and $14°$; what will the latter mark, when the former marks $16°$?

Saturday, *May* 31, 1862.

FIRST DIVISION—A.

1. If two weights, acting perpendicularly on a straight *Lever* on opposite sides of the fulcrum, balance each other, they are inversely as their distances from the fulcrum; and the pressure on the fulcrum is equal to their sum.

The pressure on the fulcrum is 12 lbs. and the distance of the fulcrum from the middle point of the *Lever* is $\frac{1}{12}$th of the whole length of the *Lever*, find the forces.

2. If the adjacent sides of a parallelogram represent the component forces in direction and magnitude, the diagonal will represent the *Resultant* force in direction and magnitude.

If the *Resultant* force be represented in direction and magnitude by a given straight line, give a geometrical construction for representing its two components, each of which is represented in magnitude by a given straight line. When is this impossible?

3. Describe the *Wheel-and-Axle*, and prove that there is equilibrium upon the *Wheel-and-Axle*, when the *Power* is to the *Weight* as the radius of the *Axle* to the radius of the *Wheel*.

4. The weight (W) being on an *Inclined Plane*, and the force (P) acting parallel to the plane, there is equilibrium, when $P : W ::$ the height of the plane : its length.

Find also the *pressure* on the plane.

If the pressure on the plane be equal to the force P, find the inclination of the plane, and express the pressure in terms of W.

5. Define *Velocity*.

Assuming that the arcs, which subtend equal angles at the centres of two circles, are as the radii of the circles, shew that, if P and W balance each other on the *Wheel-and-Axle*; and the whole be put in motion, $P : W :: W$'s velocity : P's velocity.

If P's velocity exceed W's velocity by 7 feet per second, and the radius of the *Wheel* be 15 times that of the *Axle*; find the velocities of P and W.

6. Define the *Centre of Gravity*.

Find the *Centre of Gravity* of two heavy points; and shew that the pressure at the *Centre of Gravity* is equal to the sum of the weights in all positions.

Find the *Centre of Gravity* of weights 2, 4, and 6, lbs. placed respectively at the angular points of a triangle.

7. The pressure upon any particle of a fluid of uniform density is proportional to its depth below the surface of the fluid.

8. Explain the *Hydrostatic Paradox*.

If a pipe, whose height above the bottom of a vessel is 112 feet, be inserted vertically into the vessel, and the whole be filled with water; find the pressure in tons on the bottom of the vessel, the area of the bottom being 4 square feet, and the weight of a cubic foot of water 1000 oz. avoirdupois.

9. If a body floats on a fluid, it displaces as much of the fluid as is equal in weight to the weight of the body.

If a body, exposed to the pressure of the air, float in water, prove that it will rise very slightly out of the water as the barometer rises, and sink a little deeper as the barometer falls.

10. When the body is immersed in fluid, the weight lost : whole weight of the body :: the *Specific Gravity* of the fluid : the *Specific Gravity* of the body.

11. Describe the construction of the *Common Pump;* and explain its mode of operation, and the conditions requisite to its working.

Which of these conditions will be affected by the height of the *Barometer?*

12. Describe the *Barometer*, and shew that the pressure of the atmosphere is accurately measured by the weight of the column of mercury in the *Barometer*.

If a *Barometer* were used by a diver under water, what change would take place in its height?

FIRST DIVISION.—B.

1. If two forces, acting perpendicularly on a straight *Lever* in opposite directions and on the same side of the fulcrum, balance each other, they are inversely as their distances from the fulcrum; and the pressure on the fulcrum is equal to the difference of the forces.

The pressure on the fulcrum is 4 lbs., and the distance of the fulcrum from the middle point of the *Lever* is twice the whole length of the *Lever;* find the forces.

2. If three forces, represented in magnitude and direction by the sides of a triangle, act on a point, they will keep it at rest.

Three forces whose magnitudes are 6, 8, and 10, lbs. respectively, acting upon a point, keep it at rest, prove that the directions of two of the forces are at right angles to each other.

3. In a system, in which the same string passes round any number of pulleys, and the parts of it between the pulleys are parallel, there is equilibrium, when power (P) : weight (W) :: 1 : the number of strings at the lower block.

If 6 strings *pass* round the lower block, and a man supports himself by standing on it and holding the rope which passes round the pulleys with his hands; find the *tension* of the rope.

4. The weight (W) being on an *Inclined Plane*, and the force (P) acting parallel to the plane, there is equilibrium, when $P : W ::$ the height of the plane : its length.

Find also the *pressure* on the plane.

If the pressure : power :: 3 : 4, express each of them in terms of W.

5. Define *Velocity*, and prove that if P and W balance each other in the system described in the preceding question and the whole be put in motion,

$$P : W :: W\text{'s velocity} : P\text{'s velocity}.$$

If P descends 12 feet, while W ascends through 2 feet, find the number of strings at the lower block.

6. Find the *Centre of Gravity* of a triangle.

Shew that, if the *Centre of Gravity* of three heavy particles placed at the angular points of the triangle coincides with the *Centre of Gravity* of the triangle, the particles must be of equal weight.

7. Define a *Fluid*, and distinguish between elastic and non-elastic fluids. Describe any experiment to shew that fluids press equally in all directions.

8. The surface of every fluid at rest is horizontal.

Explain clearly why this proposition is not true for very large surfaces of water.

9. If a body floats on a fluid, it displaces as much of the fluid as is equal in weight to the weight of the body.

If a body float partly immersed in two or more fluids, state the conditions of equilibrium.

If a body, floating in water and exposed to the atmospheric pressure, be placed under an exhausted *Receiver*, shew that the body will sink a little deeper.

10. Define *Specific Gravity*; and shew how to determine by means of the *Hydrostatic Balance* the *Specific Gravity* of a substance specifically heavier than water.

11. Describe the construction of the *Common Air-pump* and its operation.

How is the degree of exhaustion limited in this pump?

10—2

12. Describe the *Siphon* and explain its mode of action, and the conditions requisite to its working.

Which of these conditions will be affected by the height of the *Barometer?*

SECOND DIVISION.—A.

1. Define the terms *Force* and *Weight;* and explain the term *Mass* of a body.

Why is the *Weight* of the same body not constant at all points of the earth's surface? Could this difference in weight be detected by a *Common Balance?*

2. If two weights, acting perpendicularly on a straight *Lever* on opposite sides of the fulcrum, balance each other, they are inversely as their distances from the fulcrum; and the pressure on the fulcrum is equal to their sum.

If the *Lever* is in equilibrium, when weights P and Q are suspended from its extremities, and also when P is doubled and Q increased by 5 lbs.; find the magnitude of Q.

3. Explain the meaning of *component* and *resultant* forces.

Enunciate the *Parallelogram of Forces;* and prove the proposition, so far as the *direction* of the *Resultant* is concerned.

If a given force be resolved into two equal component forces, prove that the extremities of these forces always lie on a certain straight line.

4. In a system, in which each pulley hangs by a separate string, and the strings are parallel, there is equilibrium, when $P : W :: 1 :$ that power of 2 whose index is the number of moveable pulleys.

Supposing that there are n moveable pulleys, and that the weight of each of them is P, what must be the value of W, when there is equilibrium?

5. Define *Velocity*, and prove that, if a weight W be supported on an *Inclined Plane* by a power P acting parallel to the plane, and the whole be set in motion,

$$P : W :: W\text{'s velocity} : P\text{'s velocity}.$$

6. When a body is placed on a horizontal plane, it will stand or fall, according as the vertical line, drawn from its *Centre of Gravity*, falls within or without its base.

Shew that a rhombus will rest in stable equilibrium, when any side is placed on a horizontal plane.

7. Explain how fluid *pressures* are measured, taking as an illustration any familiar instance.

If the pressure of the atmosphere be 14 lbs. on the square inch, what will be the pressure in tons on the square yard?

8. The pressure upon any particle of a fluid of uniform density is proportional to its depth below the surface of the fluid.

Water floats on mercury to the depth of 17 feet; compare with the atmospheric pressure the pressure at a point 15 inches below the surface of the mercury, taking into account the atmospheric pressure on the surface of the water, having given that the heights of the mercurial and water-barometers are 30 inches and 34 feet respectively.

9. Define *Specific Gravity;* and prove that when a body of uniform density floats on a fluid, the part immersed : the whole body :: the *Specific Gravity* of the body : the *Specific Gravity* of the fluid.

10. The elastic force of air at a given temperature varies as the *Density*.

11. Describe the *Common Air-pump*, and its operation; and explain why a perfect vacuum cannot be obtained by this instrument.

12. Explain the method of filling and graduating a common mercurial *Thermometer*.

If the difference of readings of a *Thermometer*, which is graduated both according to Fahrenheit's and the Centigrade scale, be 40, find the temperature in each scale.

SECOND DIVISION.—B.

1. Define the terms *Force* and *Weight;* and enumerate the chief forces with which we have to deal in *Statics*.

Explain the mode of representing and comparing *Forces*.

2. If two forces, acting perpendicularly on a straight *Lever* in opposite directions and on the same side of the fulcrum, balance each other, they are inversely as their distances from the fulcrum; and the pressure on the fulcrum is equal to the difference of the forces.

The *Lever* is in equilibrium under the action of the forces P and Q, and is also in equilibrium when P is trebled, and Q increased by 6 lbs: find the magnitude of Q.

3. Explain the meaning of *Resultant* force.

Enunciate the *Parallelogram of Forces;* and assuming its truth so far as the *direction* of the *Resultant* is concerned, prove the proposition completely.

Find a point within a quadrilateral, such that, if forces be represented by lines drawn from it to the angular points of the quadrilateral, the forces shall be in equilibrium.

4. Describe the *Wheel-and-Axle;* and prove there is equilibrium upon the *Wheel-and-Axle*, when the *Power* is to the *Weight* as the radius of the *Axle* to the radius of the *Wheel*.

5. Define *Velocity*, and prove that if P and W balance in that system of pulleys, in which each pulley hangs by a separate string and the strings are parallel, and the whole be put in motion,

$$P : W :: W\text{'s velocity} : P\text{'s velocity}.$$

6. Define the term *Centre of Gravity;* and prove that when a body is suspended from a point, it will rest with its *Centre of Gravity* in the vertical line passing through the point of suspension.

How may we employ this proposition to determine practically the *Centre of Gravity* of any body whatsoever?

7. Define a *Fluid*, and distinguish between elastic and non-elastic fluids. Describe an experiment to prove that fluid pressures are transmitted equally in all directions.

8. If a body floats on a fluid, it displaces as much of the fluid as is equal in weight to the weight of the body; and it presses downwards, and is pressed upwards, with a force equal to the weight of the fluid displaced.

What further condition is necessary for equilibrium?

Two bodies of unequal *volume*, when weighed in air, balance one another; what will take place if they be weighed (1) in vacuo, (2) when the *Barometer* rises?

9. When a body is immersed in fluid, the weight lost : whole weight of the body :: the *Specific Gravity* of the fluid : the *Specific Gravity* of the body.

Why will this proposition not be true, if the fluid be compressible?

10. Shew that air has weight; and explain clearly why a balloon ascends. Why does it cease to ascend?

11. Describe the construction of the *Common Mercurial Barometer*, and prove, that the pressure of the atmosphere is accurately measured by the weight of the column of mercury in the *Barometer*.

12. Explain the action of the *Siphon*, and the conditions under which it works.

A *Siphon*, filled with water, has its ends inserted in vessels filled with water; state what will take place when the vertical distances of the highest point of the *Siphon* above the surfaces of the fluid are both less, both greater, and one greater and the other less, than the height of the *Water Barometer*.

MONDAY, *June* 1, 1863.

(*A*).

1. If two forces, acting perpendicularly upon a straight *Lever* in opposite directions and on the same side of the fulcrum, balance each other, they are inversely proportional to their distances from the fulcrum; and the pressure on the fulcrum is equal to the difference of the forces.

A uniform heavy rod AB (8 inches long) is laid over two props (C and D), which are 5 inches apart, and have their extremities in the same horizontal line, so that A is distant 2 inches from C. Find the pressures on the props; and shew that, if a weight be laid upon the rod at B sufficient to remove all pressure from the prop C, the pressure on D will be *ten times* as great as it was before.

2. Enunciate the *Parallelogram of Forces*. Assuming it to be true for the *direction* of the *Resultant*, prove it for the *magnitude* of the *Resultant*.

If the straight lines AB, AC, meeting in a point, represent two forces in direction and magnitude, the straight line joining A with the middle point of BC will represent half the *Resultant* of the two forces.

3. In a system of moveable pulleys, in which each pulley hangs by a separate string and the strings are parallel, there is equilibrium where
$$P : W :: 1 : 2^n,$$
the weights of the pulleys being neglected.

If there are three pulleys of equal weight, what must be the weight of each, in order that a weight of 56 pounds attached to the lowest may be supported by a power equal to 7 lbs. 14 oz.?

4. On the *Inclined Plane*, where the power acts parallel to the plane, prove that

$P : W :: W$'s velocity in the direction of gravity : P's velocity.

What is meant by saying that in a machine "what is gained in power is lost in time"?

5. Mention any causes which may make the weight of a body vary when it is estimated at different places.

Define the *Centre of Gravity* of a body. Find that of a plane triangle.

6. When a body is placed upon a horizontal plane, it will stand or fall according as the vertical line, drawn from its *Centre of Gravity*, falls within or without the base.

7. Define a *Fluid*. Describe an experiment by which it may be shewn that fluids press equally in all directions.

A closed vessel full of fluid, with its upper surface horizontal, has a weak part in its upper surface not capable of bearing a pressure of more than $4\frac{1}{2}$ pounds on the square foot. If a piston, the area of which is 2 square inches, be fitted into an aperture in the upper surface, what pressure applied to it will burst the vessel?

8. The surface of a fluid at rest is a horizontal plane.

9. If a body float in a fluid, it displaces as much fluid as is equal in weight to the weight of the body.

10. Explain what is meant by the *Density*, and by the *Specific Gravity*, of a body. What is meant by saying that the *Specific Gravity* of a substance is 1·5 for instance?

UNIVERSITY EXAMINATION PAPERS. 153

Prove the relation existing between the *Weight, Magnitude,* and *Specific Gravity,* of a body.

Two metals are combined into a lump, the volume of which is 2 cubic inches. $1\frac{1}{2}$ cubic inches of one metal weigh as much as the lump, and $2\frac{1}{2}$ cubic inches of the other metal weigh the same. What *Volume* of each of the two metals is there in the lump?

11. State the law connecting the elastic force and the density of air at a given temperature.

Describe the *Barometer*.

If the weight of the column of mercury, which is above the exposed surface be an ounce, and the area of the transverse section of the tube $\frac{1}{224}$ of a square inch, what is the pressure of the atmosphere on a square inch?

12. Describe the action of the *Siphon*.

(B).

1. If two weights, acting perpendicularly upon a straight *Lever* on opposite sides of the fulcrum, balance each other, they are inversely as their distances from the fulcrum; and the pressure on the fulcrum is equal to their sum.

A body, the weight of which is one pound, when placed in one scale of a false balance, appears to weigh 14 ounces. What will be its apparent weight, when placed in the other scale?

2. Enunciate the *Parallelogram of Forces*. Prove it so far as the *direction* of the *Resultant* is concerned.

Two forces, one of which is double of the other, act upon a point, and are such, that, if 6 lbs. be added to the larger, and the smaller be doubled, the direction of the *Resultant* is unaltered. Find the forces.

3. When a weight (W) is supported upon a smooth *Inclined Plane* by a power (P) acting parallel to the plane, then

$$P : W :: \text{height of the plane} : \text{its length}.$$

A weight W is supported upon an *Inclined Plane* by a string parallel to its length. The string passes over a fixed pulley, and then under a moveable one without weight, to which a weight W is attached, and having the portions of the string on each side of it parallel. Prove that the height of the plane is half its length.

4. In the system of pulleys, in which each pulley hangs by a separate string, prove that

$$P : W :: W\text{'s velocity} : P\text{'s velocity.}$$

What is meant by saying that in mechanism "what is gained in power is lost in time"?

5. Mention any causes, which may affect the weight of a body, when it is estimated at different places.

Define the *Centre of Gravity* of a body.

Shew how to find that of any number of heavy particles.

6. When a body is suspended from a point, about which it can swing freely, it will rest with its *Centre of Gravity* in the vertical line passing through the point of suspension.

7. Define a fluid. Describe an experiment, by which it may be shewn, that fluids press equally in all directions.

A closed vessel full of fluid, with its upper surface horizontal, has a weak part in its upper surface, not capable of bearing a pressure of more than 9 lbs. upon the square foot. If a piston, the area of which is one square inch, be fitted into an aperture in the upper surface, what pressure applied to it will burst the vessel?

8. Explain what is meant by the *Density*, and *Specific Gravity*, of a body. What do you mean by saying that the density of a substance is 1·5 for instance?

Prove the relation existing between the *Mass*, *Volume*, and *Density*, of a body.

Two metals are combined into a lump, the volume of which is 3 cubic inches. $2\frac{1}{2}$ cubic inches of one metal weigh as much as the lump, and $3\frac{1}{2}$ cubic inches of the other metal weigh the same. What *Volume* of each of the two metals is there in the lump?

9. The pressure at any point of a fluid of uniform density at rest is proportional to the depth below the surface of the fluid.

10. When a body is immersed in a fluid the weight lost : whole weight of the body :: the *Specific Gravity* of the fluid : the *Specific Gravity* of the body.

11. State the law connecting the elastic force and the density of air at a given temperature.

Describe the *Barometer*.

If the weight of the column of mercury, which is above the exposed surface, be an ounce, and the area of the transverse section of the tube be $\frac{1}{224}$ of a square inch, what is the pressure of the atmosphere upon a square inch?

12. Describe the action of the *Siphon*.

THE END.

Cambridge:
PRINTED BY C. J. CLAY, M.A.
AT THE UNIVERSITY PRESS.

BY THE SAME AUTHOR.

1. THE NECESSITY OF A STUDIOUS AND LEARNED CLERGY. *A Visitation Sermon.* 1*s.*

2. THE TRUE THEORY OF MISSIONS, according to Holy Scripture. *A Sermon.* 1*s.*

3. THE READING AND DEPORTMENT OF THE CLERGY DURING DIVINE SERVICE. 6*d.*

4. WOOD'S ALGEBRA, much enlarged, with numerous Problems and Examples. *Sixteenth Edition.* 12*s.* 6*d. boards.*

5. A SHORT AND EASY COURSE OF ALGEBRA, with Easy Exercises. *Sixth Edition.* 2*s.* 6*d. boards.*

6. A KEY TO THE SHORT AND EASY COURSE OF ALGEBRA, containing Solutions of all the *Exercises.* Stitched. 2*s.* 6*d.*

7. ELEMENTS OF GEOMETRY AND MENSURATION, with Easy Exercises, designed for Schools and Adult Classes, in Three Parts.

   Part I.—Geometry as a Science. 2*s.*

   Part II.—Geometry as an Art. 2*s.*

   Part III.—Geometry combined with Arithmetic. (*Mensuration.*) 3*s.* 6*d.* KEY, by the Rev. F. Calder, 3*s.*

   Parts I. and II. may be had together in one volume, *boards*, price 3*s.* 6*d.*, forming a short comprehensive treatise on *Geometry*, both theoretical and practical; and the three parts together, in *boards*, price 7*s.*

   "Mr Lund's '*Geometry as a Science*' contains, in 87 pages, the most useful propositions of the first six books of Euclid. * * * For

Schools and adult classes it is peculiarly and happily adapted, and we earnestly recommend it as a very useful little volume."—*English Journal of Education.*

"A better introduction (Part I.) to Geometry can hardly be imagined; and we are glad to find it is published at so reasonable a price, because it is a work that ought to be used extensively in our national schools of design, and widely circulated among our manufacturing population."—*Athenæum.*

8. COMPANION TO WOOD'S ALGEBRA (for *Students*), containing Solutions of all the *difficult* Questions and Problems in the Algebra. *Third Edition,* 7s. 6d. *boards.*

9. A KEY for *Schoolmasters* to all the Questions and Problems in WOOD'S ALGEBRA, by Lund. 7s. 6d. *boards.*

10. GEOMETRICAL EXERCISES, with Solutions, forming a KEY to Parts I. and II. of the GEOMETRY. Price 3s. 6d. This work contains a numerous collection of original easy 'RIDERS' to EUCLID, adapted to the Senate-House Examinations for B.A. Degree.

11. A KEY TO BISHOP COLENSO'S BIBLICAL ARITHMETIC. *Second Edition,* 1s.

LONDON:
LONGMAN, GREEN, LONGMAN, ROBERTS, AND GREEN.

By J. C. SNOWBALL, M.A.

LATE FELLOW OF ST JOHN'S COLLEGE, CAMBRIDGE.

PLANE AND SPHERICAL TRIGONOMETRY,

WITH

THE CONSTRUCTION AND USE OF TABLES OF LOGARITHMS.

Tenth Edition. 240 pp. (1863). Crown 8vo. 7*s.* 6*d.*

In preparing a new edition, the proofs of some of the more important propositions have been rendered more strict and general; and a considerable addition of more than *Two hundred Examples*, taken principally from the questions in the Examinations of Colleges and the University, has been made to the collection of Examples and Problems for practice.

MACMILLAN AND CO.
London and Cambridge.

AN ELEMENTARY LATIN GRAMMAR,

By H. J. ROBY, M.A.

UNDER-MASTER OF DULWICH COLLEGE UPPER SCHOOL,
LATE FELLOW AND CLASSICAL LECTURER OF ST JOHN'S COLLEGE,
CAMBRIDGE.

18mo. 2s. 6d.

The Author's experience in practical teaching has induced an attempt to treat Latin Grammar in a more precise and intelligible way than has been usual in school books. The facts have been derived from the best authorities, especially Madvig's Grammar and other works. The works also of Lachmann, Ritschl, Key, and others have been consulted on special points. The accidence and prosody have been simplified and restricted to what is really required by boys. In the Syntax an analysis of sentences has been given, and the uses of the different cases, tenses and moods briefly but carefully described. Particular attention has been paid to a classification of the uses of the subjunctive mood, to the prepositions, the *oratio obliqua*, and such sentences as are introduced by the English 'that.' Appendices treat of the Latin forms of Greek nouns, abbreviations, dates, money, &c. The Grammar is written in English.

MACMILLAN AND CO.
London and Cambridge.

May, 1869.

LIST OF EDUCATIONAL BOOKS

PUBLISHED BY

MACMILLAN AND CO.,

16, *BEDFORD STREET, COVENT GARDEN,*

𝔏𝔬𝔫𝔡𝔬𝔫, w.c.

CONTENTS.

	Page
CLASSICAL	3
MATHEMATICAL	7
SCIENCE	17
MISCELLANEOUS	18
DIVINITY	21
BOOKS ON EDUCATION	24

Messrs. MACMILLAN & Co. *beg to call attention to the accompanying Catalogue of their* EDUCATIONAL WORKS, *the writers of which are mostly scholars of eminence in the Universities, as well as of large experience in teaching.*

Many of the works have already attained a wide circulation in England and in the Colonies, and are acknowledged to be among the very best Educational Books on their respective subjects.

The books can generally be procured by ordering them through local booksellers in town or country, but if at any time difficulty should arise, Messrs. MACMILLAN *will feel much obliged by direct communication with themselves on the subject.*

Notices of errors or defects in any of these works will be gratefully received and acknowledged.

LIST OF EDUCATIONAL BOOKS.

CLASSICAL.

ÆSCHYLI EUMENIDES. The Greek Text, with English Notes, and English Verse Translation and an Introduction. By BERNARD DRAKE, M.A., late Fellow of King's College, Cambridge. 8vo. 7s. 6d.

> The Greek Text adopted in this Edition is based upon that of Wellauer, which may be said in general terms to represent that of the best manuscripts. But in correcting the Text, and in the Notes, advantage has been taken of the suggestions of Hermann, Paley, Linwood, and other commentators.

ARISTOTLE ON FALLACIES; OR, THE SOPHISTICI ELENCHI. With a Translation and Notes by EDWARD POSTE, M.A., Fellow of Oriel College, Oxford. 8vo. 8s. 6d.

> Besides the doctrine of Fallacies, Aristotle offers either in this treatise, or in other passages quoted in the commentary, various glances over the world of science and opinion, various suggestions on problems which are still agitated, and a vivid picture of the ancient system of dialectics, which it is hoped may be found both interesting and instructive.
>
> " It is not only scholarlike and careful; it is also perspicuous."—*Guardian.*

ARISTOTLE.—AN INTRODUCTION TO ARISTOTLE'S RHETORIC. With Analysis, Notes, and Appendices. By E. M. COPE, Senior Fellow and Tutor of Trinity College, Cambridge. 8vo. 14s.

> This work is introductory to an edition of the Greek Text of Aristotle's Rhetoric, which is in course of preparation.
>
> " Mr. Cope has given a very useful appendage to the promised Greek Text; but also a work of so much independent use that he is quite justified in his separate publication. All who have the Greek Text will find themselves supplied with a comment; and those who have not will find an analysis of the work."—*Athenæum.*

CATULLI VERONENSIS LIBER, edited by R. ELLIS, Fellow of Trinity College, Oxford. 18mo. 3s. 6d.

> " It is little to say that no edition of Catullus at once so scholarlike has ever appeared in England."—*Athenæum.*
>
> " Rarely have we read a classic author with so reliable, acute, and safe a guide."—*Saturday Review.*

CICERO.—THE SECOND PHILIPPIC ORATION. With an Introduction and Notes, translated from the German of KARL HALM. Edited, with Corrections and Additions, by JOHN E. B. MAYOR, M.A., Fellow and Classical Lecturer of St. John's College, Cambridge. Third Edition, revised. Fcap. 8vo. 5*s*.

> "A very valuable edition, from which the student may gather much both in the way of information directly communicated, and directions to other sources of knowledge."—*Athenæum.*

DEMOSTHENES ON THE CROWN. The Greek Text with English Notes. By B. DRAKE, M.A., late Fellow of King's College, Cambridge. Third Edition, to which is prefixed ÆSCHINES AGAINST CTESIPHON, with English Notes. Fcap. 8vo. 5*s*.

> The terseness and felicity of Mr. Drake's translations constitute perhaps the chief value of his edition, and the historical and archæological details necessary to understanding the *De Corona* have in some measure been anticipated in the notes on the Oration of Æschines. In both, the text adopted in the Zurich edition of 1851, and taken from the Parisian MS., has been adhered to without any variation. Where the readings of Bekker, Dissen, and others appear preferable, they are subjoined in the notes.

HODGSON.—MYTHOLOGY FOR LATIN VERSIFICATION. A Brief Sketch of the Fables of the Ancients, prepared to be rendered into Latin Verse for Schools. By F. HODGSON, B.D., late Provost of Eton. New Edition, revised by F. C. HODGSON, M.A. 18mo. 3*s*.

> Intending the little book to be entirely elementary, the Author has made it as easy as he could, without too largely superseding the use of the Dictionary and Gradus. By the facilities here afforded, it will be possible, in many cases, for a boy to get rapidly through these preparatory exercises; and thus, having mastered the first difficulties, he may advance with better hopes of improvement to subjects of higher character, and verses of more difficult composition.

JUVENAL, FOR SCHOOLS. With English Notes. By J. E. B. MAYOR, M.A. New and Cheaper Edition. Crown 8vo.

[In the Press.

> "A School edition of Juvenal, which, for really ripe scholarship, extensive acquaintance with Latin literature, and familiar knowledge of Continental criticism, ancient and modern, is unsurpassed, we do not say among English School-books, but among English editions generally."—*Edinburgh Review.*

LYTTELTON.—THE COMUS of MILTON rendered into Greek Verse. By LORD LYTTELTON. Extra fcap. 8vo. Second Edition. 5*s*.

— THE SAMSON AGONISTES of MILTON rendered into Greek Verse. By LORD LYTTELTON. Extra fcap. 8vo. 6*s*. 6*d*.

CLASSICAL.

MARSHALL.—A TABLE OF IRREGULAR GREEK VERBS, Classified according to the Arrangement of Curtius's Greek Grammar. By J. M. MARSHALL, M.A., Fellow and late Lecturer of Brasenose College, Oxford; one of the Masters in Clifton College. 8vo. cloth. 1s.

MAYOR.—FIRST GREEK READER. Edited after KARL HALM, with Corrections and large Additions by the Rev. JOHN E. B. MAYOR, M.A., Fellow and Classical Lecturer of St. John's College, Cambridge. Fcap. 8vo. 6s.

MAYOR.—GREEK for BEGINNERS. By the Rev. JOSEPH B. MAYOR, M.A. With Glossary and Index. Fcap. 8vo. 4s. 6d.

MERIVALE.—KEATS' HYPERION rendered into Latin Verse. By C. MERIVALE, B.D. Second Edition. Extra fcap. 8vo. 3s. 6d.

PLATO.—THE REPUBLIC OF PLATO. Translated into English, with an Analysis and Notes, by J. Ll. DAVIES, M.A., and D. J. VAUGHAN, M.A. Third Edition, with Vignette Portraits of Plato and Socrates, engraved by JEENS from an Antique Gem. 18mo. 4s. 6d.

ROBY.—A LATIN GRAMMAR for the Higher Classes in Grammar Schools. By H. J. ROBY, M.A.; based on the "Elementary Latin Grammar." [*In the Press.*

SALLUST.—CAII SALLUSTII CRISPI Catilina et Jugurtha. For use in Schools (with copious Notes). By C. MERIVALE, B.D. (In the present Edition the Notes have been carefully revised, and a few remarks and explanations added.) Second Edition. Fcap. 8vo. 4s. 6d.

The Jugurtha and the Catilina may be had separately, price 2s. 6d. each.

TACITUS.—THE HISTORY OF TACITUS translated into ENGLISH. By A. J. CHURCH, M.A., and W. J. BRODRIBB, M.A. With Notes and a Map. 8vo. 10s. 6d.

> "Every classical student will do well to purchase this excellent translation." —*Educational Times.*
>
> "It furnishes a very close translation, in thoroughly readable idiomatic English."—*British Quarterly Review.*

— THE AGRICOLA and GERMANY. By the same translators. With Maps and Notes. Extra fcap. 8vo. 2s. 6d.

EDUCATIONAL BOOKS.

THRING.—Works by **Edward Thring, M.A.**, Head Master of Uppingham School :—

— A CONSTRUING BOOK. Fcap. 8vo. 2s. 6d.

> This Construing Book is drawn up on the same sort of graduated scale as the Author's *English Grammar*. Passages out of the best Latin Poets are gradually built up into their perfect shape. The few words altered, or inserted as the passages go on, are printed in Italics. It is hoped by this plan that the learner, whilst acquiring the rudiments of language, may store his mind with good poetry and a good vocabulary.

— A LATIN GRADUAL. A First Latin Construing Book for Beginners. Fcap. 8vo. 2s. 6d.

> The main plan of this little work has been well tested.
> The intention is to supply by easy steps a knowledge of Grammar, combined with a good vocabulary; in a word, a book which will not require to be forgotten again as the learner advances.
> A short practical manual of common Mood constructions, with their English equivalents, form the second part.

— A MANUAL of MOOD CONSTRUCTIONS. Extra fcap. 8vo. 1s. 6d.

THUCYDIDES.—THE SICILIAN EXPEDITION. Being Books VI. and VII. of Thucydides, with Notes. A New Edition, revised and enlarged, with a Map. By the Rev. PERCIVAL FROST, M.A., late Fellow of St. John's College, Cambridge. Fcap. 8vo. 5s.

> This edition is mainly a grammatical one. Attention is called to the force of compound verbs, and the exact meaning of the various tenses employed.

WRIGHT.—Works by **J. Wright, M.A.**, late Head Master of Sutton Coldfield School :—

— HELLENICA; Or, a HISTORY of GREECE in GREEK, as related by Diodorus and Thucydides, being a First Greek Reading Book, with Explanatory Notes Critical and Historical. Third Edition, with a Vocabulary. 12mo. 3s. 6d.

> In the last twenty chapters of this volume, Thucydides sketches the rise and progress of the Athenian Empire in so clear a style and in such simple language, that the author doubts whether any easier or more instructive passages can be selected for the use of the pupil who is commencing Greek.

— A HELP TO LATIN GRAMMAR; Or, the Form and Use of Words in Latin, with Progressive Exercises. Crown 8vo. 4s. 6d.

> "Never was there a better aid offered alike to teacher and scholar in that arduous pass. The style is at once familiar and strikingly simple and lucid; and the explanations precisely hit the difficulties, and thoroughly explain them."—*English Journal of Education.*

WRIGHT.—Works by **J. Wright, M.A.**—*Continued.*

— THE SEVEN KINGS OF ROME. An Easy Narrative, abridged from the First Book of Livy by the omission of difficult passages, being a First Latin Reading Book, with Grammatical Notes. Fcap. 8vo. 3*s.*

> This work is intended to supply the pupil with an easy Construing-book, which may at the same time be made the vehicle for instructing him in the rules of grammar and principles of composition. Here Livy tells his own pleasant stories in his own pleasant words. Let Livy be the master to teach a boy Latin, not some English collector of sentences, and he will not be found a dull one.

— A VOCABULARY AND EXERCISES on the "SEVEN KINGS OF ROME." Fcap. 8vo. 2*s.* 6*d.*

The Vocabulary and Exercises may also be had bound up with "The Seven Kings of Rome," price 5*s.*

MATHEMATICAL.

AIRY.—Works by **G. B. Airy,** Astronomer Royal :—

— ELEMENTARY TREATISE ON PARTIAL DIFFERENTIAL EQUATIONS. Designed for the use of Students in the University. With Diagrams. Crown 8vo. cloth, 5*s.* 6*d.*

> It is hoped that the methods of solution here explained, and the instances exhibited, will be found sufficient for application to nearly all the important problems of Physical Science, which require for their complete investigation the aid of partial differential equations.

— ON THE ALGEBRAICAL AND NUMERICAL THEORY of ERRORS of OBSERVATIONS, and the COMBINATION of OBSERVATIONS. Crown 8vo. cloth, 6*s.* 6*d.*

— UNDULATORY THEORY OF OPTICS. Designed for the use of Students in the University. New Edition. Crown 8vo. cloth, 6*s.* 6*d.*

— ON SOUND and ATMOSPHERIC VIBRATIONS. With the Mathematical Elements of Music. Designed for the use of Students of the University. Crown 8vo. 9*s.*

BAYMA.—THE ELEMENTS of MOLECULAR MECHANICS. By JOSEPH BAYMA, S.J., Professor of Philosophy, Stonyhurst College. Demy 8vo. cloth, 10*s.* 6*d.*

BOOLE.—Works by **G. Boole, D.C.L., F.R.S.**, Professor of Mathematics in the Queen's University, Ireland :—

— A TREATISE ON DIFFERENTIAL EQUATIONS. New and Revised Edition. Edited by I. TODHUNTER. Crown 8vo. cloth, 14s.

> The author has endeavoured in this Treatise to convey as complete an account of the present state of knowledge on the subject of Differential Equations, as was consistent with the idea of a work intended primarily for elementary instruction. The earlier sections of each chapter contain that kind of matter which has usually been thought suitable to the beginner, while the later ones are devoted either to an account of recent discovery, or the discussion of such deeper questions of principle as are likely to present themselves to the reflective student in connexion with the methods and processes of his previous course.

— A TREATISE ON DIFFERENTIAL EQUATIONS. Supplementary Volume. Edited by I. TODHUNTER. Crown 8vo. cloth, 8s. 6d.

— THE CALCULUS OF FINITE DIFFERENCES. Crown 8vo. cloth, 10s. 6d.

> This work is in some measure designed as a sequel to the *Treatise on Differential Equations*, and is composed on the same plan.

BEASLEY.—AN ELEMENTARY TREATISE ON PLANE TRIGONOMETRY. With Examples. By R. D. BEASLEY, M.A., Head Master of Grantham Grammar School. Second Edition, revised and enlarged. Crown 8vo. cloth, 3s. 6d.

> This Treatise is specially intended for use in Schools. The choice of matter has been chiefly guided by the requirements of the three days' Examination at Cambridge, with the exception of proportional parts in Logarithms, which have been omitted. About *Four hundred* Examples have been added, mainly collected from the Examination Papers of the last ten years, and great pains have been taken to exclude from the body of the work any which might dishearten a beginner by their difficulty.

CAMBRIDGE SENATE-HOUSE PROBLEMS and RIDERS, WITH SOLUTIONS :—

1848—1851.—PROBLEMS. By FERRERS and JACKSON. 8vo. cloth. 15s. 6d.

1848—1851.—RIDERS. By JAMESON. 8vo. cloth. 7s. 6d.

1854.—PROBLEMS and RIDERS. By WALTON and MACKENZIE, 8vo. cloth. 10s. 6d.

1857.—PROBLEMS and RIDERS. By CAMPION and WALTON. 8vo. cloth. 8s. 6d.

1860.—PROBLEMS and RIDERS. By WATSON and ROUTH. Crown 8vo. cloth. 7s. 6d.

1864.—PROBLEMS and RIDERS. By WALTON and WILKINSON. 8vo. cloth. 10s. 6d.

CAMBRIDGE COURSE OF ELEMENTARY NATURAL PHILOSOPHY, for the Degree of B.A. Originally compiled by J. C. SNOWBALL, M.A., late Fellow of St. John's College. Fifth Edition, revised and enlarged, and adapted for the Middle-Class Examinations by THOMAS LUND, B.D., Late Fellow and Lecturer of St. John's College; Editor of Wood's Algebra, &c. Crown 8vo. cloth. 5s.

CAMBRIDGE AND DUBLIN MATHEMATICAL JOURNAL. THE COMPLETE WORK, in Nine Vols. 8vo. cloth. £7 4s.
(Only a few copies remain on hand.)

CANDLER.—HELP to ARITHMETIC. Designed for the use of Schools. By H. CANDLER, M.A., Mathematical Master at Uppingham. Fcap. 8vo. 2s. 6d.

CHEYNE.—AN ELEMENTARY TREATISE on the PLANETARY THEORY. With a Collection of Problems. By C. H. H. CHEYNE, B.A. Crown 8vo. cloth. 6s. 6d.

— THE EARTH'S MOTION of ROTATION. By C. H. H. CHEYNE, M.A. Crown 8vo. 3s. 6d.

CHILDE.—THE SINGULAR PROPERTIES of the ELLIPSOID and ASSOCIATED SURFACES of the Nth DEGREE. By the Rev. G. F. CHILDE, M.A., Author of "Ray Surfaces," "Related Caustics," &c. 8vo. 10s. 6d.

CHRISTIE.—A COLLECTION OF ELEMENTARY TEST-QUESTIONS in PURE and MIXED MATHEMATICS; with Answers and Appendices on Synthetic Division, and on the Solution of Numerical Equations by Horner's Method. By JAMES R. CHRISTIE, F.R.S., late First Mathematical Master at the Royal Military Academy, Woolwich. Crown 8vo. cloth, 8s. 6d.

DALTON.—ARITHMETICAL EXAMPLES. Progressively arranged, with Exercises and Examination Papers. By the Rev. T. DALTON, M.A., Assistant Master of Eton College. 18mo. cloth. 2s. 6d.

DAY.—PROPERTIES OF CONIC SECTIONS PROVED GEOMETRICALLY. Part I., THE ELLIPSE, with Problems. By the Rev. H. G. DAY, M.A., Head Master of Sedbergh Grammar School. Crown 8vo. 3s. 6d.

DODGSON.—AN ELEMENTARY TREATISE ON DETERMINANTS, with their Application to Simultaneous Linear Equations and Algebraical Geometry. By C. L. DODGSON, M.A., Mathematical Lecturer of Christ Church, Oxford. Small 4to. cloth, 10s. 6d.

DREW.—GEOMETRICAL TREATISE on CONIC SECTIONS. By W. H. DREW, M.A., St. John's College, Cambridge. Third Edition. Crown 8vo. cloth, 4s. 6d.

> In this work the subject of Conic Sections has been placed before the student in such a form that, it is hoped, after mastering the elements of Euclid, he may find it an easy and interesting continuation of his geometrical studies. With a view also of rendering the work a complete Manual of what is required at the Universities, there have been either embodied into the text, or inserted among the examples, every book-work question, problem, and rider, which has been proposed in the Cambridge examinations up to the present time.

— SOLUTIONS TO THE PROBLEMS IN DREW'S CONIC SECTIONS. Crown 8vo. cloth, 4s. 6d.

FERRERS.—AN ELEMENTARY TREATISE on TRILINEAR CO-ORDINATES, the Method of Reciprocal Polars, and the Theory of Projections. By the Rev. N. M. FERRERS, M.A., Fellow and Tutor of Gonville and Caius College, Cambridge. Second Edition. Crown 8vo. 6s. 6d.

> The object of the author in writing on this subject has mainly been to place it on a basis altogether independent of the ordinary Cartesian system, instead of regarding it as only a special form of Abridged Notation. A short chapter on Determinants has been introduced.

FROST.—THE FIRST THREE SECTIONS of NEWTON'S PRINCIPIA. With Notes and Illustrations. Also a Collection of Problems, principally intended as Examples of Newton's Methods. By PERCIVAL FROST, M.A., late Fellow of St. John's College, Mathematical Lecturer of King's College, Cambridge. Second Edition. 8vo. cloth, 10s. 6d.

> The author's principal intention is to explain difficulties which may be encountered by the student on first reading the Principia, and to illustrate the advantages of a careful study of the methods employed by Newton, by showing the extent to which they may be applied in the solution of problems; he has also endeavoured to give assistance to the student who is engaged in the study of the higher branches of Mathematics, by representing in a geometrical form several of the processes employed in the Differential and Integral Calculus, and in the analytical investigations of Dynamics.

FROST and WOLSTENHOLME.—A TREATISE ON SOLID GEOMETRY. By PERCIVAL FROST, M.A., and the Rev. J. WOLSTENHOLME, M.A., Fellow and Assistant Tutor of Christ's College. 8vo. cloth, 18s.

> The authors have endeavoured to present before students as comprehensive a view of the subject as possible. Intending as they have done to make the subject accessible, at least in the earlier portion, to all classes of students, they have endeavoured to explain fully all the processes which are most useful in dealing with ordinary theorems and problems, thus directing the student to the selection of methods which are best adapted to the exigencies of each problem. In the more difficult portions of the subject, they have considered themselves to be addressing a higher class of students; there they have tried to lay a good foundation on which to build, if any reader should wish to pursue the science beyond the limits to which the work extends.

GODFRAY.—A TREATISE on ASTRONOMY, for the use of Colleges and Schools. By HUGH GODFRAY, M.A., Mathematical Lecturer at Pembroke College, Cambridge. 8vo. cloth. 12s. 6d.

> "We can recommend for its purpose a very good *Treatise on Astronomy* by Mr. Godfray. It is a working book, taking astronomy in its proper place in mathematical science. But it begins with the elementary definitions, and connects the mathematical formulæ very clearly with the visible aspect of the heavens and the instruments which are used for observing it."—*Guardian.*

— AN ELEMENTARY TREATISE on the LUNAR THEORY. With a brief Sketch of the Problem up to the time of Newton. By HUGH GODFRAY, M.A. Second Edition, revised. Crown 8vo. cloth. 5s. 6d.

HEMMING.—AN ELEMENTARY TREATISE on the DIFFERENTIAL AND INTEGRAL CALCULUS, for the use of Colleges and Schools. By G. W. HEMMING, M.A., Fellow of St. John's College, Cambridge. Second Edition, with Corrections and Additions. 8vo. cloth. 9s.

JONES and CHEYNE.—ALGEBRAICAL EXERCISES. Progressively arranged. By the Rev. C. A. JONES, M.A., and C. H. CHEYNE, M.A., Mathematical Masters of Westminster School. New Edition. 18mo. cloth, 2s. 6d.

> This little book is intended to meet a difficulty which is probably felt more or less by all engaged in teaching Algebra to beginners. It is that while new ideas are being acquired, old ones are forgotten. In the belief that constant practice is the only remedy for this, the present series of miscellaneous exercises has been prepared. Their peculiarity consists in this, that though miscellaneous they are yet progressive, and may be used by the pupil almost from the commencement of his studies. They are not intended to supersede the systematically arranged examples to be found in ordinary treatises on Algebra, but rather to supplement them.
> The book being intended chiefly for Schools and Junior Students, the higher parts of Algebra have not been included.

KITCHENER.—A GEOMETRICAL NOTE-BOOK, containing Easy Problems in Geometrical Drawing preparatory to the Study of Geometry. For the use of Schools. By F. E. KITCHENER, M.A., Mathematical Master at Rugby. 4to. 2s.

MORGAN.—A COLLECTION of PROBLEMS and EXAMPLES in Mathematics. With Answers. By H. A. MORGAN, M.A., Sadlerian and Mathematical Lecturer of Jesus College, Cambridge. Crown 8vo. cloth. 6s. 6d.

> This book contains a number of problems, chiefly elementary, in the Mathematical subjects usually read at Cambridge. They have been selected from the papers set during late years at Jesus college. Very few of them are to be met with in other collections, and by far the larger number are due to some of the most distinguished Mathematicians in the University.

PARKINSON.—Works by **S. Parkinson, B.D.**, Fellow and Prælector of St. John's College, Cambridge:—

— AN ELEMENTARY TREATISE ON MECHANICS. For the use of the Junior Classes at the University and the Higher Classes in Schools. With a Collection of Examples. Third Edition, revised. Crown 8vo. cloth, 9s. 6d.

> The author has endeavoured to render the present volume suitable as a Manual for the junior classes in Universities and the higher classes in Schools. In the Third Edition several additional propositions have been incorporated in the work for the purpose of rendering it more complete, and the Collection of Examples and Problems has been largely increased.

— A TREATISE on OPTICS. Second Edition, revised. Crown 8vo. cloth, 10s. 6d.

> A collection of Examples and Problems has been appended to this work which are sufficiently numerous and varied in character to afford useful exercise for the student: for the greater part of them recourse has been had to the Examination Papers set in the University and the several Colleges during the last twenty years.

PHEAR.—ELEMENTARY HYDROSTATICS. With numerous Examples. By J. B. PHEAR, M.A., Fellow and late Assistant Tutor of Clare College, Cambridge. Fourth Edition. Crown 8vo. cloth, 5s. 6d.

> "An excellent Introductory Book. The definitions are very clear; the descriptions and explanations are sufficiently full and intelligible; the investigations are simple and scientific. The examples greatly enhance its value."—*English Journal of Education.*

PRATT.—A TREATISE on ATTRACTIONS, LAPLACE'S FUNCTIONS, and the FIGURE of the EARTH. By JOHN H. PRATT, M.A., Archdeacon of Calcutta, Author of "The Mathematical Principles of Mechanical Philosophy." Third Edition. Crown 8vo. cloth, 6s. 6d.

PUCKLE.—AN ELEMENTARY TREATISE on CONIC SECTIONS and ALGEBRAIC GEOMETRY. With numerous Examples and hints for their Solution; especially designed for the use of Beginners. By G. H. PUCKLE, M.A., St. John's College, Cambridge, Head Master of Windermere College. Third Edition, enlarged and improved. Crown 8vo. cloth, 7s. 6d.

> The work has been completely re-written, and a considerable amount of new matter has been added, to suit the requirements of the present time.

REYNOLDS.—MODERN METHODS IN ELEMENTARY GEOMETRY. By E. M. REYNOLDS, M.A., Mathematical Master in Clifton College. Crown 8vo. 3s. 6d.

RAWLINSON.—ELEMENTARY STATICS. By G. RAWLINSON, M.A. Edited by EDWARD STURGES, M.A., of Emmanuel College, Cambridge, and late Professor of the Applied Sciences, Elphinstone College, Bombay. Crown 8vo. cloth. 4*s.* 6*d.*

> Published under the authority of H. M. Secretary of State for use in the Government Schools and Colleges in India.
> "This Manual may take its place among the most exhaustive, yet clear and simple, we have met with, upon the composition and resolution of forces, equilibrium, and the mechanical powers."—*Oriental Budget.*

ROUTH.—AN ELEMENTARY TREATISE on the DYNAMICS of a SYSTEM of RIGID BODIES. With Examples. By EDWARD JOHN ROUTH, M.A., Fellow and Assistant Tutor of St. Peter's College, Cambridge; Examiner in the University of London. Second Edition. Crown 8vo. cloth, 14*s.*

SMITH.—A TREATISE on ELEMENTARY STATICS. By J. H. SMITH, M.A., Gonville and Caius College, Cambridge. 8vo. 5*s.* 6*d.*

— A TREATISE on ELEMENTARY HYDROSTATICS. By J. H. SMITH, M.A. 8vo. 4*s.* 6*d.*

— A TREATISE on ELEMENTARY TRIGONOMETRY. By J. H. SMITH, M.A. 8vo. 5*s.*

SMITH.—Works by **Barnard Smith, M.A.**, Rector of Glaston, Rutlandshire, late Fellow and Senior Bursar of St. Peter's College, Cambridge:—

— ARITHMETIC and ALGEBRA, in their Principles and Application, with numerous Systematically arranged Examples, taken from the Cambridge Examination Papers, with especial reference to the Ordinary Examination for B.A. Degree. Tenth Edition. Crown 8vo. cloth, 10*s.* 6*d.*

> This work is now extensively used in *Schools* and *Colleges* both *at home* and in the *Colonies*. It has also been found of great service for students preparing for the MIDDLE-CLASS AND CIVIL AND MILITARY SERVICE EXAMINATIONS, from the care that has been taken to elucidate the *principles* of all the Rules.

— ARITHMETIC FOR SCHOOLS. New Edition. Crown 8vo. cloth, 4*s.* 6*d.*

— COMPANION to ARITHMETIC for SCHOOLS. [*Preparing.*

— A KEY to the ARITHMETIC for SCHOOLS. Seventh Edition. Crown 8vo., cloth, 8*s.* 6*d.*

— EXERCISES in ARITHMETIC. With Answers. Crown 8vo. limp cloth, 2*s.* 6*d.* Or sold separately, as follows:—Part I. 1*s.*; Part II. 1*s.* ANSWERS, 6*d.*

> These Exercises have been published in order to give the pupil examples in every rule of Arithmetic. The greater number have been carefully compiled from the latest University and School Examination Papers.

SMITH.—Works by **Barnard Smith, M.A.**—*Continued.*

— SCHOOL CLASS-BOOK of ARITHMETIC. 18mo. cloth, 3s. Or sold separately, Parts I. and II. 10d. each; Part III. 1s.

— KEYS to SCHOOL CLASS-BOOK of ARITHMETIC. Complete in one Volume, 18mo., cloth, 6s. 6d.; or Parts I., II., and III. 2s. 6d. each.

— SHILLING BOOK of ARITHMETIC for NATIONAL and ELEMENTARY SCHOOLS. 18mo. cloth. Or separately, Part I. 2d.; Part II. 3d.; Part III. 7d. ANSWERS, 6d.

THE SAME, with Answers complete. 18mo. cloth, 1s. 6d.

— KEY to SHILLING BOOK of ARITHMETIC. 18mo. cloth, 4s. 6d.

— EXAMINATION PAPERS in ARITHMETIC. In Four Parts. 18mo. cloth, 1s. 6d. THE SAME, with Answers, 18mo. 1s. 9d.

— KEY to EXAMINATION PAPERS in ARITHMETIC. 18mo. cloth, 4s. 6d.

SNOWBALL.—PLANE and SPHERICAL TRIGONOMETRY. With the Construction and Use of Tables of Logarithms. By J. C. SNOWBALL. Tenth Edition. Crown 8vo. cloth, 7s. 6d.

TAIT and STEELE.—DYNAMICS of a PARTICLE. With Examples. By Professor TAIT and Mr. STEELE. New Edition. Crown 8vo. cloth, 10s. 6d.

> In this Treatise will be found all the ordinary propositions connected with the Dynamics of Particles which can be conveniently deduced without the use of D'Alembert's Principles. Throughout the book will be found a number of illustrative Examples introduced in the text, and for the most part completely worked out; others, with occasional solutions or hints to assist the student, are appended to each Chapter.

TAYLOR.—GEOMETRICAL CONICS; including Anharmonic Ratio and Projection, with numerous Examples. By C. TAYLOR, B.A., Scholar of St. John's College, Cambridge. Crown 8vo. cloth, 7s. 6d.

TEBAY.—ELEMENTARY MENSURATION for SCHOOLS. With numerous Examples. By SEPTIMUS TEBAY, B.A., Head Master of Queen Elizabeth's Grammar School, Rivington. Extra fcap. 8vo. 3s. 6d.

TODHUNTER.—Works by **I. Todhunter, M.A., F.R.S.**, Fellow and Principal Mathematical Lecturer of St. John's College, Cambridge :—

— THE ELEMENTS of EUCLID for the use of COLLEGES and SCHOOLS. New Edition. 18mo. cloth, 3s. 6d.

— ALGEBRA for BEGINNERS. With numerous Examples. New Edition. 18mo. cloth, 2s. 6d.

— KEY to ALGEBRA for BEGINNERS. Crown 8vo., cl., 6s. 6d.

— TRIGONOMETRY for BEGINNERS. With numerous Examples. New Edition. 18mo. cloth, 2s. 6d.

> Intended to serve as an introduction to the larger treatise on *Plane Trigonometry*, published by the author. The same plan has been adopted as in the *Algebra for Beginners*: the subject is discussed in short chapters, and a collection of examples is attached to each chapter.

— MECHANICS for BEGINNERS. With numerous Examples. 18mo. cloth, 4s. 6d.

> Intended as a companion to the two preceding books. The work forms an elementary treatise on *Demonstrative* Mechanics. It may be true that this part of mixed mathematics has been sometimes made too abstract and speculative; but it can hardly be doubted that a knowledge of the elements at least of the theory of the subject is extremely valuable even for those who are mainly concerned with practical results. The author has accordingly endeavoured to provide a suitable introduction to the study of applied as well as of theoretical Mechanics.

— A TREATISE on the DIFFERENTIAL CALCULUS. With Examples. Fourth Edition. Crown 8vo. cloth, 10s. 6d.

— A TREATISE on the INTEGRAL CALCULUS. Third Edition, revised and enlarged. With Examples. Crown 8vo. cloth, 10s. 6d.

— A TREATISE on ANALYTICAL STATICS. With Examples. Third Edition, revised and enlarged. Crown 8vo. cloth, 10s. 6d.

— PLANE CO-ORDINATE GEOMETRY, as applied to the Straight Line and the CONIC SECTIONS. With numerous Examples. Fourth Edition. Crown 8vo. cloth, 7s. 6d.

— ALGEBRA. For the use of Colleges and Schools. Fourth Edition. Crown 8vo. cloth, 7s. 6d.

> This work contains all the propositions which are usually included in elementary treatises on Algebra, and a large number of *Examples for Exercise*. The author has sought to render the work easily intelligible to students without impairing the accuracy of the demonstrations, or contracting the limits of the subject. The Examples have been selected with a view to illustrate every part of the subject, and as the number of them is about *Sixteen hundred and fifty*, it is hoped they will supply ample exercise for the student. Each set of Examples has been carefully arranged, commencing with very simple exercises, and proceeding gradually to those which are less obvious.

TODHUNTER.—Works by **I. Todhunter, M.A.**—*Continued.*

— PLANE TRIGONOMETRY. For Schools and Colleges. Third Edition. Crown 8vo. cloth, 5s.

> The design of this work has been to render the subject intelligible to beginners, and at the same time to afford the student the opportunity of obtaining all the information which he will require on this branch of Mathematics. Each chapter is followed by a set of Examples; those which are entitled *Miscellaneous Examples*, together with a few in some of the other sets, may be advantageously reserved by the student for exercise after he has made some progress in the subject. In the Second Edition the hints for the solution of the Examples have been considerably increased.

— A TREATISE ON SPHERICAL TRIGONOMETRY. Second Edition, enlarged. Crown 8vo. cloth, 4s. 6d.

> This work is constructed on the same plan as the *Treatise on Plane Trigonometry*, to which it is intended as a sequel. Considerable labour has been expended on the text in order to render it comprehensive and accurate, and the Examples, which have been chiefly selected from University and College Papers, have all been carefully verified.

— EXAMPLES of ANALYTICAL GEOMETRY of THREE DIMENSIONS. Second Edition, revised. Crown 8vo. cloth, 4s.

— AN ELEMENTARY TREATISE on the THEORY of EQUATIONS. Second Edit., revised. Cr. 8vo. cl., 7s. 6d.

WILSON.—ELEMENTARY GEOMETRY. PART I. Angles, Parallels, Triangles, and Equivalent Figures, with the Application to Problems. By J. M. WILSON, M.A., Fellow of St. John's College, Cambridge, and Mathematical Master in Rugby School. Extra fcap. 8vo. 2s. 6d.

> "It is an actual substitute for the first two books of Euclid, in which many of his propositions are drawn out from the conception of straightness, parallelism, angles, with wonderful ease and simplicity, and the methods employed have the great merit of suggesting a ready application to the solution of fresh problems."—*Guardian.*

——ELEMENTARY GEOMETRY, PART II. THE CIRCLE AND PROPORTION. By J. M. WILSON, M.A. Extra fcap. 8vo. 2s. 6d.

— A TREATISE on DYNAMICS. By W. P. WILSON, M.A., Fellow of St. John's College, Cambridge; and Professor of Mathematics in Queen's College, Belfast. 8vo. 9s. 6d.

WOLSTENHOLME.—A BOOK of MATHEMATICAL PROBLEMS on subjects included in the Cambridge Course. By JOSEPH WOLSTENHOLME, Fellow of Christ's College, sometime Fellow of St. John's College, and lately Lecturer in Mathematics at Christ's College. Crown 8vo. cloth. 8s. 6d.

> CONTENTS: Geometry (Euclid).—Algebra.—Plane Trigonometry.—Conic Sections, Geometrical.—Conic Sections, Analytical.—Theory of Equations. —Differential Calculus.—Integral Calculus.—Solid Geometry.—Statics.— Dynamics, Elementary.—Newton.—Dynamics of a Point.—Dynamics of a Rigid Body.—Hydrostatics.—Geometrical Optics.—Spherical Trigonometry and Plane Astronomy.

SCIENCE.

AIRY.—POPULAR ASTRONOMY. With Illustrations. By G. B. AIRY, Astronomer Royal. Sixth and Cheaper Edition. 18mo. cloth, 4s. 6d.

> "Popular Astronomy in general has many manuals; but none of them supersede the Six Lectures of the Astronomer Royal under that title. Its speciality is the direct way in which every step is referred to the observatory, and in which the methods and instruments by which every observation is made are fully described. This gives a sense of solidity and substance to astronomical statements which is obtainable in no other way."—*Guardian.*

GEIKIE.—ELEMENTARY LESSONS in PHYSICAL GEOLOGY. By ARCHIBALD GEIKIE, F.R.S., Director of the Geological Survey of Scotland. [*Preparing.*

HUXLEY.—LESSONS in ELEMENTARY PHYSIOLOGY. With numerous Illustrations. By T. H. HUXLEY, F.R.S., Professor of Natural History in the Royal School of Mines. Seventh Thousand. 18mo. cloth, 4s. 6d.

> "It is a very small book, but pure gold throughout. There is not a waste sentence, or a superfluous word, and yet it is all clear as daylight. It exacts close attention from the reader, but the attention will be repaid by a real acquisition of knowledge. And though the book is so small, it manages to touch on some of the very highest problems. . . . . The whole book shows how true it is that the most elementary instruction is best given by the highest masters in any science."—*Guardian.*

> "The very best descriptions and explanations of the principles of human physiology which have yet been written by an Englishman."—*Saturday Review.*

QUESTIONS on HUXLEY'S PHYSIOLOGY for SCHOOLS. By T. ALCOCK, M.D. 18mo. 1s. 6d.

LOCKYER.—ELEMENTARY LESSONS in ASTRONOMY. With Coloured Diagram of the Spectra of the Sun, Stars, and Nebulæ, and numerous Illustrations. By J. NORMAN LOCKYER, F.R.A.S. 18mo. 5s. 6d.

> "Forms an admirable introduction to the study of astronomy. While divested of anything like irrelevant disquisition it is popular in its method of treatment, and written in clear and easily comprehended language."—*Educational Times.*

> "It is remarkably clear and compact, the illustrations are also of unusual excellence. No other book on the subject that we know is at once so small and so good."—*Guardian.*

OLIVER.—LESSONS IN ELEMENTARY BOTANY. With nearly Two Hundred Illustrations. By DANIEL OLIVER, F.R.S., F.L.S. Fourth Thousand. 18mo. cloth, 4s. 6d.

"The manner is most fascinating, and if it does not succeed in making this division of science interesting to every one, we do not think anything can. . . . . Nearly 200 well executed woodcuts are scattered through the text, and a valuable and copious index completes a volume which we cannot praise too highly, and which we trust all our botanical readers, young and old, will possess themselves of."—*Popular Science Review.*

"To this system we now wish to direct the attention of teachers, feeling satisfied that by some such course alone can any substantial knowledge of plants be conveyed with certainty to young men educated as the mass of our medical students have been. We know of no work so well suited to direct the botanical pupil's efforts as that of Professor Oliver's, who, with views so practical and with great knowledge too, can write so accurately and clearly."—*Natural History Review.*

ROSCOE.—LESSONS in ELEMENTARY CHEMISTRY, Inorganic and Organic. By HENRY ROSCOE, F.R.S., Professor of Chemistry in Owen's College, Manchester. With numerous Illustrations and Chromo-Litho. of the Solar Spectra. Twelfth Thousand. 18mo. cloth, 4s. 6d.

It has been the endeavour of the author to arrange the most important facts and principles of Modern Chemistry in a plain but concise and scientific form, suited to the present requirements of elementary instruction. For the purpose of facilitating the attainment of exactitude in the knowledge of the subject, a series of exercises and questions upon the lessons have been added. The metric system of weights and measures, and the centigrade thermometric scale, are used throughout the work.

"A small, compact, carefully elaborated and well arranged manual."—*Spectator.*

MISCELLANEOUS.

ATLAS of EUROPE. GLOBE EDITION. Uniform in size with Macmillan's Globe Series, containing 48 Coloured Maps, on the same scale Plans of London and Paris, and a copious Index, strongly bound in half-morocco, with flexible back. 9s.

NOTICE.—This Atlas includes all the Countries of Europe in a Series of Forty-eight Maps, drawn on the same scale, with an Alphabetical Index to the situation of more than 10,000 Places; and the relation of the various Maps and Countries to each other is defined in a general Key-Map.

"In the series of works which Messrs. Macmillan and Co. are publishing under this general title *(Globe Series)* they have combined portableness with scholarly accuracy and typographical beauty, to a degree that is almost unprecedented. Happily they are not alone in employing the highest available scholarship in the preparation of the most elementary educational works; but their exquisite taste and large resources secure an artistic result which puts them almost beyond competition. This little atlas will be an invaluable boon for the school, the desk, or the traveller's portmanteau."—*British Quarterly Review.*

BATES and LOCKYER.—A CLASS BOOK of GEOGRAPHY, adapted to the recent Programme of the Royal Geographical Society. By H. W. BATES and J. N. LOCKYER, F.R.A.S.
[*In the Press.*

CAMEOS from ENGLISH HISTORY. From Rollo to Edward II. By the Author of "The Heir of Redclyffe." Extra fcap. 8vo. 5s.

> "They are a series of vivid pictures which will not easily fade from the minds of the young people for whom they are written."—*Guardian.*
> "An admirable school book. We know of no elementary history that combines, in an equal degree, accurate knowledge with the skilful presentation of it."—*British Quarterly Review.*
> "Instead of dry details we have living pictures, faithful, vivid, and striking."—*Nonconformist.*

EARLY EGYPTIAN HISTORY for the Young. With Descriptions of the Tombs and Monuments. New Edition, with Frontispiece. Fcap. 8vo. 5s.

> "Artistic appreciation of the picturesque, lively humour, unusual aptitude for handling the childish intellect, a pleasant style, and sufficient learning, altogether free from pedantic parade, are among the good qualities of this volume, which we cordially recommend to the parents of inquiring and book-loving boys and girls."—*Athenæum.*
> "This is one of the most perfect books for the young that we have ever seen. The author has hit the best possible way of interesting any one, young or old."—*Literary Churchman.*

HISTORICAL SELECTIONS. Readings from the best Authorities on English and European History. Selected and Arranged by E. M. SEWELL and C. M. YONGE. Crown 8vo. 6s.

> "General histories are apt to be dry and meagre: but particular periods or subjects have been treated brilliantly and attractively by different authors. If these could be made to tell, by well-selected extracts, a continuous, or nearly continuous, tale, the advantage would obviously be great. This is what Miss Sewell and Miss Yonge have attempted in the volume before us. The extracts are well chosen, the volume is exceedingly interesting, and the superiority. both in the communication of permanent knowledge, and in the discipline of taste, which it possesses over all ordinary school histories, is enormous. We know of scarcely anything which is so likely to raise to a higher level the average standard of English education."—*The Guardian.*

HOLE.—A GENEALOGICAL STEMMA of the KINGS of ENGLAND and FRANCE. By the Rev. C. HOLE. On Sheet. 1s.

— **A BRIEF BIOGRAPHICAL DICTIONARY.** Compiled and Arranged by CHARLES HOLE, M.A., Trinity College, Cambridge. Second Edition, 18mo., neatly and strongly bound in cloth, 4s. 6d.

> The most comprehensive Biographical Dictionary in English,—containing more than 18,000 names of persons of all countries, with dates of birth and death, and what they were distinguished for.
> "An invaluable addition to our manuals of reference, and from its moderate price, it cannot fail to become as popular as it is useful."—*Times.*

HOUSEHOLD (A) BOOK OF ENGLISH POETRY. Selected and arranged, with Notes, by R. C. TRENCH, D.D., Archbishop of Dublin. Extra fcap. 8vo. 5s. 6d.

> "The Archbishop has conferred in this delightful volume an important gift on the whole English-speaking population of the world."—*Pall Mall Gazette.*
>
> "Remarkable for the number of fine poems it contains that are not found in other collections."—*Express.*
>
> "The selection is made with the most refined taste, and with excellent judgment."—*Birmingham Gazette.*

JEPHSON.—SHAKESPEARE'S TEMPEST. With Glossary and Explanatory Notes. By the Rev. J. M. JEPHSON. 18mo. 1s. 6d.

OPPEN.—FRENCH READER. For the use of Colleges and Schools. Containing a Graduated Selection from Modern Authors in Prose and Verse; and copious Notes, chiefly Etymological. By EDWARD A. OPPEN. Fcap. 8vo. cloth, 4s. 6d.

A SHILLING BOOK of GOLDEN DEEDS. A Reading-Book for Schools and General Readers. By the Author of "The Heir of Redclyffe." 18mo. cloth.

> "To collect in a small handy volume some of the most conspicuous of these (examples) told in a graphic and spirited style, was a happy idea, and the result is a little book that we are sure will be in almost constant demand in the parochial libraries and schools for which it is avowedly intended."—*Educational Times.*

A SHILLING BOOK of WORDS from the POETS. By C. M. VAUGHAN. 18mo. cloth.

THRING.—Works by **Edward Thring, M.A.**, Head Master of Uppingham:—

— THE ELEMENTS of GRAMMAR taught in ENGLISH. With Questions. Fourth Edition. 18mo. 2s.

— THE CHILD'S GRAMMAR. Being the substance of "The Elements of Grammar taught in English," adapted for the use of Junior Classes. A New Edition. 18mo. 1s.

> The author's effort in these two books has been to point out the broad, beaten, every-day path, carefully avoiding digressions into the bye-ways and eccentricities of language. This work took its rise from questionings in National Schools, and the whole of the first part is merely the writing out in order the answers to questions which have been used already with success. Its success, not only in National Schools, from practical work in which it took its rise, but also in classical schools, is full of encouragement.

— SCHOOL SONGS. A collection of Songs for Schools. With the Music arranged for Four Voices. Edited by the Rev. E. THRING and H. RICCIUS. Folio. 7s. 6d.

DIVINITY.

EASTWOOD.—THE BIBLE WORD BOOK. A Glossary of Old English Bible Words. By J. EASTWOOD, M.A., of St. John's College, and W. ALDIS WRIGHT, M.A., Trinity College, Cambridge. 18mo. 5s. 6d.

HARDWICK.—A HISTORY of the CHRISTIAN CHURCH. MIDDLE AGE. From Gregory the Great to the Excommunication of Luther. By ARCHDEACON HARDWICK. Edited by FRANCIS PROCTER, M.A. With Four Maps constructed for this work by A. KEITH JOHNSTON. Second Edition. Crown 8vo. 10s. 6d.

— A HISTORY of the CHRISTIAN CHURCH during the REFORMATION. By ARCHDEACON HARDWICK. Revised by FRANCIS PROCTER, M.A. Second Edition. Crown 8vo. 10s. 6d.

MACLEAR.—Works by the **Rev. G. F. Maclear, B.D.**, Head Master of King's College School, and Preacher at the Temple Church :—

— A CLASS-BOOK of OLD TESTAMENT HISTORY. Fourth Edition, with Four Maps. 18mo. cloth, 4s. 6d.

> "A work which for fulness and accuracy of information may be confidently recommended to teachers as one of the best text-books of Scripture History which can be put into a pupil's hands."—*Educational Times.*

— A CLASS-BOOK of NEW TESTAMENT HISTORY: including the Connection of the Old and New Testament. With Four Maps. Second Edition. 18mo. cloth. 5s. 6d.

> "Mr. Maclear has produced in this handy little volume a singularly clear and orderly arrangement of the Sacred Story.... His work is solidly and completely done."—*Athenæum.*

— A SHILLING BOOK of OLD TESTAMENT HISTORY, for National and Elementary Schools. With Map. 18mo. cloth.

— A SHILLING BOOK of NEW TESTAMENT HISTORY, for National and Elementary Schools. With Map. 18mo. cloth.

MACLEAR.—Works by **Rev. G. F. Maclear, B.D.**—*Continued.*

— CLASS BOOK of the CATECHISM of the CHURCH of ENGLAND. Second Edition. 18mo. cloth, 2s. 6d.

> "It is indeed the work of a scholar and divine, and as such though extremely simple it is also extremely instructive. There are few clergy who would not find it useful in preparing candidates for confirmation; and there are not a few who would find it useful to themselves as well."—*Literary Churchman.*

— A FIRST CLASS-BOOK of the CATECHISM of the CHURCH of ENGLAND, with Scripture Proofs, for Junior Classes and Schools. 18mo. 6d.

— THE ORDER OF CONFIRMATION. A Sequel to the CLASS-BOOK OF THE CHURCH CATECHISM, comprising—The Service of Confirmation, Explanation, Notes, Texts, and suitable Devotions. 18mo. 3d.

PROCTER.—A HISTORY of the BOOK of COMMON PRAYER: with a Rationale of its Offices. By FRANCIS PROCTER, M.A. Seventh Edition, revised and enlarged. Crown 8vo. 10s. 6d.

> In the course of the last twenty years the whole question of Liturgical knowledge has been reopened with great learning and accurate research, and it is mainly with the view of epitomizing their extensive publications, and correcting by their help the errors and misconceptions which had obtained currency, that the present volume has been put together.

PROCTER and MACLEAR.—AN ELEMENTARY INTRODUCTION to the BOOK of COMMON PRAYER. Third Edition, re-arranged and supplemented by an Explanation of the Morning and Evening Prayer and the Litany. By the Rev. F. PROCTER and the Rev. G. F. MACLEAR. 18mo. 2s. 6d.

PSALMS of DAVID Chronologically Arranged. By FOUR FRIENDS. An amended version, with Historical Introduction and Explanatory Notes. Crown 8vo., 10s. 6d.

> "It is a work of choice scholarship and rare delicacy of touch and feeling."—*British Quarterly.*

RAMSAY.—THE CATECHISER'S MANUAL; or, the Church Catechism illustrated and explained, for the use of Clergymen, Schoolmasters, and Teachers. By ARTHUR RAMSAY, M.A. Second Edition. 18mo. 1s. 6d.

SIMPSON.—AN EPITOME of the HISTORY of the CHRISTIAN CHURCH. By WILLIAM SIMPSON, M.A. Fourth Edition. Fcap. 8vo. 3s. 6d.

SWAINSON.—A HAND-BOOK to BUTLER'S ANALOGY. By C. A. SWAINSON, D.D., Norrisian Professor of Divinity at Cambridge. Crown 8vo. 1s. 6d.

WESTCOTT.—A GENERAL SURVEY of the HISTORY of the CANON of the NEW TESTAMENT during the First Four Centuries. By BROOKE FOSS WESTCOTT, B.D., Assistant Master at Harrow. Second Edition, revised. Crown 8vo. 10s. 6d.

> The Author has endeavoured to connect the history of the New Testament Canon with the growth and consolidation of the Church, and to point out the relation existing between the amount of evidence for the authenticity of its component parts and the whole mass of Christian literature. Such a method of inquiry will convey both the truest notion of the connexion of the written Word with the living Body of Christ, and the surest conviction of its divine authority.
>
> "Mr. Westcott's 'Introduction to the Study of the Gospels,' and his 'History of the Canon' are two of the best works of the kind to be found in any literature, and exhibit the solidity of English judgment in combination with a fulness of learning which is often assumed to be a monopoly of the Germans."—*Daily News.*

— INTRODUCTION to the STUDY of the FOUR GOSPELS. By BROOKE FOSS WESTCOTT, B.D. Third Edition. Crown 8vo. 10s. 6d.

> This book is intended to be an Introduction to the *Study* of the Gospels. In a subject which involves so vast a literature much must have been overlooked; but the author has made it a point at least to study the researches of the great writers, and consciously to neglect none.

— A GENERAL VIEW of the HISTORY of the ENGLISH BIBLE. By BROOKE FOSS WESTCOTT, B.D. Crown 8vo. 10s. 6d.

> "The first trustworthy account we have had of that unique and marvellous monument of the piety of our ancestors."—*Daily News.*

— THE BIBLE in the CHURCH. A Popular Account of the Collection and Reception of the Holy Scriptures in the Christian Churches. Second Edition. By BROOKE FOSS WESTCOTT, B.D. 18mo. cloth, 4s. 6d.

— THE GOSPEL of the RESURRECTION. Thoughts on its Relation to Reason and History. By BROOKE FOSS WESTCOTT, B.D. New Edition. Fcap. 8vo. 4s. 6d.

WILSON.—AN ENGLISH HEBREW and CHALDEE LEXICON and CONCORDANCE to the more Correct Understanding of the English translation of the Old Testament, by reference to the Original Hebrew. By WILLIAM WILSON, D.D., Canon of Winchester, late Fellow of Queen's College, Oxford. Second Edition, carefully Revised. 4to. cloth, 25s.

> The aim of this work is, that it should be useful to Clergymen and all persons engaged in the study of the Bible, even when they do not possess a knowledge of Hebrew; while able Hebrew scholars have borne testimony to the help that they themselves have found in it.

BOOKS ON EDUCATION.

ARNOLD.—A FRENCH ETON; or, Middle-Class Education and the State. By MATTHEW ARNOLD. Fcap. 8vo. cloth. 2s. 6d.

"A very interesting dissertation on the system of secondary instruction in France, and on the advisability of copying the system in England."—*Saturday Review.*

— SCHOOLS and UNIVERSITIES on the CONTINENT. 8vo. 10s. 6d.

BLAKE.—A VISIT to some AMERICAN SCHOOLS and COLLEGES. By SOPHIA JEX BLAKE. Crown 8vo. cloth. 6s.

"Miss Blake gives a living picture of the schools and colleges themselves, in which that education is carried on."—*Pall-Mall Gazette.*

ESSAYS ON A LIBERAL EDUCATION. By CHARLES STUART PARKER, M.A., HENRY SIDGWICK, M.A., LORD HOUGHTON, JOHN SEELEY, M.A., REV. F. W. FARRAR, M.A., F.R.S., &c., E. E. BOWEN, M.A., F.R.A.S., J. W. HALES, M.A., J. M. WILSON, M.A., F.G.S., F.R.A.S., W. JOHNSON, M.A. Edited by the Rev. F. W. FARRAR, M.A., F.R.S., late Fellow of Trinity College, Cambridge; Fellow of King's College, London; Assistant Master at Harrow; Author of "Chapters on Language," &c., &c. Second Edition. 8vo. cloth, 10s. 6d.

FARRAR.—ON SOME DEFECTS IN PUBLIC SCHOOL EDUCATION. A Lecture delivered at the Royal Institution. With Notes and Appendices. Crown 8vo. 1s.

THRING.—EDUCATION AND SCHOOL. By the Rev. EDWARD THRING, M.A., Head Master of Uppingham. Second Edition. Crown 8vo. cloth. 6s.

YOUMANS.—MODERN CULTURE: its True Aims and Requirements. A Series of Addresses and Arguments on the Claims of Scientific Education. Edited by EDWARD L. YOUMANS, M.D. Crown 8vo. 8s. 6d.

HIATUS: THE VOID in MODERN EDUCATION, its CAUSE and ANTIDOTE. By OUTIS. 8vo. 8s. 6d.

CAMBRIDGE:—PRINTED BY JONATHAN PALMER.

www.ingramcontent.com/pod-product-compliance
Lightning Source LLC
Chambersburg PA
CBHW020842160426
43192CB00007B/752